How to Do
Ecology

How to Do
Ecology

A Concise Handbook
SECOND EDITION

Richard Karban
Mikaela Huntzinger
Ian S. Pearse

Princeton University Press

Princeton and Oxford

press.princeton.edu

Cover art by Richard Karban, from a photo by
Kaori Shiojiri; Sagehen Creek, California

ISBN 978–0-691–16176–1

Library of Congress Control Number: 2014934363

British Library Cataloging-in-Publication
Data is available

This book has been composed in
ITC New Baskerville

Printed on acid-free paper ∞

Printed in the United States of America

10 9 8 7 6 5 4 3 2

Contents

Illustrations

Boxes

Preface to the Second Edition

This book started out as handouts for Rick's graduate-level field ecology course. At first there were just a few pages, but each year the stack grew thicker. Nothing happened to these handouts for about a decade until Mikaela, in the throes of graduate school herself, suggested that they might be useful to other people. She started organizing them as a fun project (seriously?!?!) and adding new material.

We had no idea who would want to read this thing. However, we now know that most of our readers are prospective and current grad students in ecology. So we have added other advice and information that we think will be valuable to students, such as how to read more efficiently, think more creatively, and find a job in ecology. In the meantime, things have changed—for example, more ecologists are using surveys and other observational techniques to learn about nature. For the second edition, Rick and Mikaela sought out Ian to include sections about model-building techniques, analysis of surveys, and ways of dealing with spatially and phylogenetically clumped observations and other design issues, and to add his perspective throughout the book.

The move from undergrad to grad school can be a shock. What matters has changed. For one thing, the currency is different. Doing well as an undergrad involved taking classes and getting good grades. In grad school, unless you are moving from a master's to a doctoral program, grades become virtually obsolete. The currency to succeed after grad school is publications and, in some cases, grants. While you are in grad school, prioritize the things that will result in publications.

The three of us hope this book makes learning to do ecology more straightforward for you than it was for us.

Introduction:
The Aims of This Book

As students of ecology, we take classes in ecological principles and ecological theory. We familiarize ourselves with the influential studies that have shaped our discipline. But rarely do we explicitly try to figure out how to do ecology ourselves. What are the skills that are required to do a good job? How can we develop them? In a nutshell, this book is an attempt to provide a concise set of suggestions for how to do ecology well. It is intended for students and practicing ecologists who are faced with developing an exciting research program.

In this handbook, we consider different ecological approaches and discuss their strengths, weaknesses, and utility. We concentrate on hypothesis testing, as this is the approach currently favored by most ecologists and the one we are most familiar with. We present some rules of thumb for how to set up experiments and how to analyze and interpret results. Many questions that ecologists are interested in answering do not lend themselves to experimental manipulation, so we also present techniques for conducting and analyzing observational surveys. Finally, we offer suggestions about working with other people, communicating what you find in scientific papers, talks, and proposals, and developing other skills that are useful for ecologists.

How to Do
Ecology

CHAPTER 1

Picking a Question

Perhaps the most critical step in doing field biology is picking a question. Tragically, it's the thing that you are expected to do first, when you have the least experience. For example, it helps to get into grad school if you appear to be focused on a particular set of questions that matches a professor's interests. However, at this stage in most students' careers, many topics sound equally interesting, so this forced focus is difficult or even painful.

The question that you pick should reflect your goals as a biologist. If you are a new grad student, your short-term goal might be nothing more than to succeed in grad school. However, it's important to look farther down the road even as you're beginning. A common mid-term goal is getting your first job. For most jobs—those at research universities, small liberal arts colleges, federal agencies, nonprofit organizations—search committees will want to see a strong record of research and publication even if you won't be expected to conduct research or publish a lot on the job. Box 1 presents a justification for this bias. Search committees want to know that you are capable of advancing the field and communicating effectively. (They may also want to see other qualifications and experiences, such as teaching.) We consider strategies for getting different kinds of jobs in ecology in chapter 7. Achieving a goal like getting a first job also demands that you build a mid-term plan for your research. For example, your plan might include solving a problem in restoration, such as how to return a particular piece of real estate to some level of ecological functioning. A more conceptual mid-term goal might involve making people rethink the interactions that are important determinants of the abundance or distribution of species.

Long-term goals are harder to formulate but are at least as important. (If you don't believe this, talk to some burnt-out researchers late in their careers. Some people never bothered to stop and figure out what they really valued and wanted to accomplish for themselves. Thinking through your big-picture, long-term goals makes doing the work more enjoyable.) Some long-term goals that you might want to try out include attempting to influence how you and others think about or practice a certain subdiscipline of biology, or how we manage a habitat or species. Such long-term goals can provide a yardstick with which to evaluate your choice of project. Your long-term goals should suit you and not necessarily your major advisor (who may consider a nonacademic career a waste of time) and not necessarily your parents (who may try to convince you that a conceptual thesis will leave you unemployable). Refer to the "How to Get a Job" section of chapter 7 to begin the difficult work of untangling your goals from theirs.

From the beginning, consider your short-, mid-, and long-term goals as you pick your research question. Push yourself to pose a question that both satisfies your goals

Box 1. *The importance of research for people who aspire to non-research careers*

Even if a career in research is not part of your long-term goals, it is still worth throwing yourself into the world of research while you work on your degree. The process of doing research will give you insights into ecology that are extremely difficult to get anywhere else.

· Doing experiments yourself helps you understand how individual biases, preconceptions, and points of view shape the ecological information that appears in textbooks.
· Over time, working on independent research helps you to incorporate scientific reasoning into your own thinking, which allows you to analyze reports and articles critically and to teach the information to others more effectively.
· Writing up your results teaches even strong writers how to write more efficiently, concisely, and clearly.

These and other insights and skills are virtually impossible to gain solely through reading; instead, you are more likely to learn these things by immersing yourself in your research. And besides, it's fun.

and will be of broad interest to others. At the same time, don't let the quest for the perfect question keep you from making tangible research progress. Figure out how narrow or broad you want your research question to be. You should recognize that if you answer a very specific question, your results may be considered important by only a very small community. Academics are more likely to get enthused about a more general question. On the other hand, it is also possible to ask a question that is too general (theoretical), so you should ask yourself if your answer will reflect reality for at least one actual species. Having a model organism in mind will keep you more grounded in reality and increase the size of your audience.

If your question is very specific, ask whether you can generalize from your results. You may find yourself answering a specific, non-conceptual question about fisheries biology, restoration, and so on if you receive funding from an applied source. It may not be possible to couch your question in more conceptual terms. If so, you may be able to ask a complementary, more general question as well. For example, your specific question might be which animals visit a particular night-blooming flower. More general (and interesting) questions might be which visitors succeed at pollinating the flower and what qualities of the flower and visitor make pollination more likely. The answer to these latter questions will be compelling to a wider audience.

Not only should your question be of broad conceptual interest, but it should also be as novel as possible. All projects have to be original to some extent. We all like to hear new stories and new ideas, and ecologists place a large premium on novelty. If you are asking the same question that has been answered in other study systems (that is, with similar organisms in analogous environments), it behooves you to think about what you can do to set your study apart from the others. That said, if you are trying to start a project and haven't yet thought of a novel question, one useful way to begin is to repeat an experiment or a study that captured your attention and imagination. Sometimes repeating a published study as a jumping off place will keep you from getting stuck and will inspire you to move in an exciting new direction.

Policy makers are much less concerned with novelty than academics are. If you are funded by an agency to answer a specific policy question, you will need to balance your academic colleagues' expectation of novelty and your funding source's demands to answer the specific question for which they are giving you money. Your first priority should be to generate relevant data for your funders; however, if possible, ask additional, complementary questions in your study system that can lead to publishable research.

So you're looking for questions that are specific yet general and novel yet relevant to your goals. You could fret about this for years. Don't obsess about thinking up the perfect study before you are willing to begin (see box 2). One of the most unsuccessful personality traits in this business is perfectionism. Field studies are never going to be perfect. For example, don't get stuck thinking that you need to read more before you can do anything else. Reading broadly is great, but you will learn more by watching, tweaking, and thinking about your system. In addition, it is not realistic to expect yourself to sit at your desk and conjure up the perfect study that will revolutionize the field. Revolutionary questions don't get asked in a vacuum; they evolve. You start asking one question, hit a few brick walls, get exposed to some ideas or observations that you hadn't previously considered, and pretty soon you're asking very different questions that are better than your initial naïve ones. Most projects don't progress as we originally conceived them.

Box 2. *Advice on picking questions
for three types of ecologists*

There are three kinds of ecologists:

· The perfectionists who can't get started,
· The jackrabbits who have a lot of energy and want to get
  started before thinking through their goals, and
· Those who are just right, someplace in between.

If you are a perfectionist who can't get started because
you haven't thought of the perfect question, we suggest you
just get out there and do it. The experience and insight
(not to mention publications) that you'll get by doing an
imperfect study will help you improve in the future. If you
are a jackrabbit and find yourself starting a million projects, our advice is to step back and ask which of these questions is most likely to advance the field and, even more
importantly, to inspire enduring passion in you. And if
you are a person who is just right, don't get a swelled head
about it.

It is fine to start by asking a relatively "small" question.
By small we mean specific to your study system and with
relatively little replication. Small questions will often generate more excitement for you than bigger ones because
their more modest goals can be achieved with relatively
few data and much more quickly. Imagine that you want to
study rates of predation on goose eggs. These eggs are difficult to find and highly seasonal. So, you could conduct a
small pilot experiment with three cartons of eggs from the
grocery store. Your pilot study will not give you definitive

answers about goose eggs but will likely provide useful insights about how to conduct that experiment. If results from the pilot study turn out as expected, they can provide a foundation for a bigger project. If the results are unexpected, they can serve as a springboard for a novel working hypothesis. Almost all of our long-term projects had their beginnings as small pilot "dabbles."

Fieldwork is a hard business, and many of the factors associated with failure or success are beyond your control. You should ask whether your ideas are feasible—are you likely to get an answer to the questions that you pose? Do you have the resources and knowledge to complete the project? To deal with the reality that field projects are hard to pull off, we suggest that you try several pilot studies simultaneously. If you know that you want to ask a particular question, try it out on several systems at the same time. You'll soon get a sense that the logistics in some systems are much more difficult than those in others, and that the biological details make some systems more amenable to answering particular questions. It is a lucky coincidence that Gregor Mendel worked on peas, since they are particularly well suited to elucidating the particulate nature of inheritance. Other people attempted to ask similar questions but were less fortunate in the systems that they chose to investigate. Since most field projects don't work, try several possibilities and follow the leads that seem the most promising. Don't get discouraged about the ones that don't work. Successful people never tell you about the many projects they didn't pull off. You should feel fortunate if two out of seven work well.

An essential ingredient of a good project is that you feel excited about it. The people who are the most successful over the long haul are those who work the hardest. No matter how disciplined you are, working hard is much easier if it doesn't feel like work but rather something that you are passionate about. You've heard the old saying, "If you have a job you love, you will never have to work a day in your life." Pick a project that is intellectually stimulating *to you*. You are the one who has to be jazzed enough about it to do the boring grunt work that all field projects involve. You will feel much more inclined to stay out there in the pouring rain, through all the mind-numbing repetitions that are required to get a large enough sample size, if you have a burning interest in your question and your system.

There are two approaches to picking a project: starting with the question or starting with the system. The difference between these two is actually smaller than it sounds because you generally have to bounce between both concerns to come out at the end with a good project. So regardless of which one you start with, you need to make sure that you are satisfying a list of criteria related to both.

Many successful studies start with a question. You may be interested in a particular kind of interaction or pattern for its own sake or because of its potential consequences. For example, you may be excited by the hypothesis that more diverse ecological systems are intrinsically more stable. Perhaps you are interested in this hypothesized relationship because if it is true, it could provide a sound rationale for conserving biodiversity, and if it is not generally true, ecologists should not attempt to use it as a basis for conservation policy. Since many studies have considered this question, you should think about what's at the bottom of the hypothesized link between biodiversity and stability. Have previous studies addressed these key elements? Are there novel aspects of this question that haven't been addressed yet? Are there assumptions that scientists take for granted but have never tested? Even questions that have been addressed by many researchers may still have components that have yet to be asked.

If you start by asking a question, you will need to find a suitable system to answer it. The system should be conveniently located. For example, if you don't have money for travel, choose a system close to home, and if you don't like to hike, choose plots near the road. Your study organisms or processes should be common enough for you to get good replication. Ideally, your sites should be protected from vandalism by curious people and animals (or it should be possible for you to minimize these risks). Your system should be amenable to the manipulations that you would like to subject it to and the observations you would like to make. You can get help finding systems by seeing what similar studies in the literature have used, by asking around, or by looking at what's available at field stations or other protected sites close to your home. The appropriate system will depend upon the specific questions that you want to ask. If your question requires you to know how your treatments affect fitness, you will want to find an annual rather than a charismatic but long-lived species. If your hypothesis relies upon a long history of coevolution, you should probably consider native systems rather than species that have been recently introduced. (Incidentally, there is a widespread chauvinism about working in pristine ecosystems. The implicit argument seems to be that the only places where we can still learn about nature are those that have not been altered by human intervention. We wonder if any such places really exist. Certainly, less disturbed places are inspiring and fun, but they also represent a very small fraction of the earth's ecosystems. There are still plenty of big questions about how nature works that can be asked in your own backyard, regardless of where you live—we can attest to this, having worked in some uninspiring places.)

One danger to guard against is trying to shoehorn a system to fit your pet hypothesis. If you start with a question, make sure you are willing to look around for the right system for that question and that you are willing to modify your question as necessary to go where the natural history of your chosen system takes you. You cannot make your organisms have a different natural history, so you must be willing to accept and work with what you encounter.

If you start with an organism or a system because of your interests, your funding, your major professor, whatever, you may find yourself in search of a question. Often one organism becomes a model for one kind of question, but it has not been explored for others. For example, the ecologies of lab darlings *Drosophila* and *Arabidopsis* are poorly known in the field. If everyone has used a system to ask one kind of question, there may be a lot of background natural history known about that system, but nobody has thought to ask the questions that you have. If you have a system but need a question, try reading broadly (and quickly) to get a sense of the kinds of questions that are exciting and interesting to you.

If you don't already have a system in mind but want to start out by taking this direction, try going to a natural area and spending a few days just looking at what's there. Generate a list of systems and questions in your notebook that you can mull over and prioritize later. Another useful approach is to start with a natural pattern you observe. First quantify that pattern. For example, you might observe that snails are at a particular density at your study site. Next ask whether there is natural variation in this measurement. Do some microhabitats have more snails than others? Is there natural variation associated with behavioral traits? For example, are the snails in some spots active but those in other places aestivating? Is there variation between individuals? Are the snails in some microenvironments bigger than those in others? Are bigger snails more active? And so on. Once you have quantified these patterns, ask more about them. What mechanisms could cause the patterns that you observe? What consequences might the patterns have on individuals and on other organisms?

Even if the pattern you observe in your scouting has been described before, there are likely to be many great projects available. If it is an important and general pattern, other people have probably noticed it too. However, it is less likely that the ecological mechanisms that cause the pattern have been evaluated. Understanding ecological mechanisms not only provides insight into how a process works, but also can tell us about its effects and where we would predict it to occur. Elucidating the mechanisms of a well-known pattern is likely to be a valuable contribution. Generate a list of potential mechanisms and then devise ways to collect evidence to test the strength of each. It is also less likely that the consequences of the pattern have been described. Does the pattern affect the fitness of the organisms that show it? Does the pattern affect their population dynamics? Does it affect the behaviors of other organisms in the system? Answering any one of these questions is plenty for a dissertation.

Don't assume that questions have been answered just because they seem obvious. For example, thousands of studies have documented predation by birds on phytophagous insects, but the effects of that predation on herbivory rates and plant fitness went relatively unexplored for decades (Marquis and Whelan 1995). More recently, effects of bird predation have been found to vary dramatically from one tree species to another (Singer et al. 2012). As another example, although periodical cicadas are the most abundant herbivores of eastern deciduous forests of North America, their interactions with their host plants and the rest of the community are largely unexplored. Pulses of dead cicada adults stimulate soil microbes and alter plant communities (Yang 2004). In short, there are still many interesting unanswered questions even in well-known systems.

Sometimes ecologists are constrained by funding sources or by labs that work on one set of organisms. If so, all of the good questions may appear to have already been addressed. Again, consider asking questions about the ecological consequences of what everyone else works on. For example, if

you work in a lab where everyone works on the morpho-
logical changes in an herbivore that are induced by expo-
sure to various predators, one more demonstration of an
induced response may not be very novel. Perhaps you can
ask what the fitness consequences of the different morphologies may be. Alternatively, try turning the question
on its head and ask how predators and competitors respond to different morphologies of the herbivores.

Once you have selected a question and collected some
preliminary data so that you know it is feasible to answer
the question, next think about how to answer it as completely as possible. One complete story will be more compelling and satisfying than a haphazard collection of loosely
related pieces. Prioritize the questions that flesh out your
best story and the questions that you can feasibly answer.
See chapter 8 for suggestions about organizing your research into one compelling story.

Here are some additional questions that could make
your study more complete.

1. Consider alternative hypotheses to produce the patterns and results that you observe (see chapter 4).
2. Think about whether the phenomenon that you
   are studying applies generally. For instance, you
   may want to repeat your studies that gave interesting results at other field sites. You might also want
   to repeat them with other species.
3. Explore whether your phenomenon operates at
   realistic spatial and temporal scales. For instance,

if you conducted a small-scale experiment, do your results apply at the larger scales where the organisms actually live (see chapter 3)?

4. If possible, work at levels both below (mechanisms) and above (consequences) the level of your pattern. What ecological mechanisms could generate the pattern that you observe? What other organisms or processes could the pattern affect?

You may not be able to answer all of these questions, but the more complete your story is, the more useful and appreciated your work is likely to be. Each of these additional questions can take a lot of time and energy, so don't expect to address them all.

Coming up with research questions can be intimidating. You'll produce better ideas if you separate idea generation from critique. Our parents, teachers, friends, society have taught us to censor our thoughts and inclinations. Failing to do so leads to so much humiliation that, as Foucault's panopticon tells us, we repress ideas that may be "incorrect" before we are even aware of them (Foucault 1977). In order to generate new ideas, we need to temporarily turn off the censor in our heads. Be willing to hang with the dumb ideas that you will inevitably come up with, because the really great ideas stand on the shoulders of the dumb ones. Creative people in all fields tend to share two traits: an ability to tolerate ambiguity (the messiness of complex problems) and a willingness to take risks and sometimes fail (Feist 1998, Martinsen 2011).

Box 3. *Generating ideas*

One technique we like is sometimes called iterative writing (Ian refers to it as "the Acid Trip"). This exercise may sound flaky, but we have found it works well for us, and even students who aren't from California have found it helpful and fun.

Before you begin, prepare for the activity:

Gather two large pieces of paper, a few different pens, and a highlighter marker.

On the first piece of paper, write a question you'd like to consider—for example, "What is my research question?" or "How should I spend this field season?" or "What do my results mean?"

Now you are ready to do the two parts that will help you generate interesting responses to the question: get relaxed and write out your ideas.

First, relax. Believe it or not, certain relaxation techniques have been shown to increase the originality and quantity of ideas (Colzato et al. 2012). We like to do a whole-body relaxation that involves consciously considering regions of the body ("Relax your toes and feet and ankles"; then "Relax your calves and your knees"; and so on).

Once you've worked your way up through your neck and face, "wake up" only your writing hand.

Now begin responding to the question you wrote at the top of the page. You are the only one who needs to see this work, so be as open to whatever comes out of your pen as possible. Remember to shut off the censor—write everything that comes into your head, whether it seems related or not. There may be times when your thoughts run dry and you can't think of anything that seems worth writing.

Box 3. Continued

We've found ourselves tempted to give up when this happens. But the process works best if you continue writing, even if you write "I don't know what to write" or rewrite something you've already written. Better yet, take a risk and write something really crazy. You'll soon find yourself generating more ideas again.

After you've done this for about 10 minutes, grab the highlighter and intuitively highlight words or phrases that seem appealing to you. Don't overthink this—keep it loose. Transfer those highlighted words and phrases to the second sheet of paper.

Use the new words and phrases to continue writing. Sometimes you will find yourself continuing to answer your original question, and other times you will go off in a new direction; it doesn't matter.

After another 10 minutes or so, repeat the process of transferring phrases and using them as a springboard for new writing if it seems productive.

When you feel like you have finished (between 20 and 40 minutes), set down your pen and slowly "wake" back up.

After you have completed this relaxation and iterative writing process, get up and take a break. When you're ready (20 minutes or two days later), go back to critique what you've written. Some of your ideas will be off the point, others will be pretty goofy, but hopefully you will have generated a few cool ideas that you otherwise wouldn't have had. These can be shared with other people, if you're so inclined. Of course, this is just the beginning, since your ideas will need to get filled out.

If you think this technique sounds good but you can't stand dealing with the handwritten output, you can do a modified version on your computer. Download a free version of OmmWriter, a program that clears your screen of everything but a vague, snowy landscape. It allows you to type but not to format, so you can't distract yourself by italicizing Latin names, or creating headings or subheadings. Just type whatever comes into your head without judging it yet. We recommend that you don't even look at the words you're typing, to avoid triggering your judgments or correcting your spelling.

We like to do another version of this technique that is a real crowd-pleaser with graduate students despite (or because of) its unconventionality. In this version, you draw out your ideas instead of writing them. Get out a large sheet of paper and your colored markers again. Write your question at the top of the paper and do a relaxation. This time, *illustrate* your responses to the question. Draw first; only use words as a last resort. Don't worry about making the pictures look real—a stick figure of the dingoes you study is fine. The key to this technique is to push yourself to draw more than you expect to. It works best if you do it well into a stage of discomfort. Keep asking, "What else can I possibly add to this?" Again, don't overthink this stage—just let the ideas keep coming out.

It is important to push criticism aside when you are generating ideas, but you'll need to come back to your writings and drawings if you want those new ideas to pay off. The next stage is to refine those ideas into a meaningful research plan.

Much of our academic training involves memorizing and critiquing the ideas of other people. But science doesn't move forward unless we generate new ideas. We are not born with creative and original thinking skills—we cultivate them. For example, we have found that we can come up with new and richer ideas by using an exercise in iterative writing that we describe in box 3. The technique encourages you to take low-stakes risks.

Another essential way to generate creative ideas is to allow your organisms to redirect your questions. Many discoveries in science are unplanned. While you are answering one question, you are likely to see things that you haven't imagined. There is some chance that nobody else has seen them either. Rather than trying to force your organisms to answer your questions, allow them to suggest new ones to you. Read broadly so that you recognize that something is novel when you stumble upon it. Above all, be opportunistic!

CHAPTER 2

Posing Questions
(or Picking an Approach)

Much of what you can learn about ecology depends on the questions you ask. Your preconceptions and intuition determine the factors that you choose to examine, and these will constrain your results. Ecologists take several different approaches to science, and which approaches you use will constrain the kinds of answers that you'll get. Answers to the questions that you ask then form your view of how the natural world works. Deciding on an approach may sound like a bunch of philosophical nonsense to waste time, but it can have important consequences on everything that follows.

Different Ways to Do Ecology

Ecologists use several different approaches to understand phenomena, which we place in three categories: (1) observations of patterns, (2) manipulative experiments, and (3) model building. As is often the case in ecology, these categories are not mutually exclusive, and each has something to offer.

Observations of Patterns or Natural History

Observations of patterns in natural systems are essential, as they provide us with the players (factors and processes) that may be important. Observations allow us to generate hypotheses and test models. Natural history used to be the mainstay in ecology, but it started to go out of style in the 1960s. Current training in ecology has become less and less based on a background in natural history (Futuyma 1998, Ricklefs 2012). Undergraduate education requires fewer hours of labs than it did in the past because labs are expensive and time-consuming to teach. Traditional courses in the "ologies" (entomology, ornithology, herpetology, etc.) are becoming endangered. Graduate students are pressured to get started on a thesis project before they've spent time poking around in real ecological systems. It doesn't get any easier for professors, who are most "successful" by becoming research administrators. They write grants to fund other people to work with the organisms, allowing themselves more time to write papers, progress reports, and the next grants. The result of all this is that the intuition for our experiments and models comes from the literature, computer models, or the intuition of our major professor. We spend a lot of time refining what everyone already believes is important. This has the danger of making ecology conservative and unexciting.

It is clear to the writers of this book that ecology as a discipline would be improved if we were encouraged to learn more about nature by observing it first and manipulating and modeling it second. Observations are absolutely necessary to provide the insights that make for good experiments and models. For example, experiments usually manipulate only one or a small number of factors because of logistical constraints. The factors that we as experimenters choose to manipulate determine the factors that we will conclude are important. For instance, if we test the hypothesis that competition affects community structure, we are more likely to learn something about competition and less likely to learn something about some other factor (such as facilitation, predation, abiotic factors, genetic structure, and so on) that we did not think to manipulate. Observe your organism or system with as few assumptions as possible and let it suggest ideas to you.

Good intuition is the first requirement for meaningful experiments. The best way to develop that intuition is by observing organisms in the field. Sadly, none of us "has the time" to spend observing nature. Guidance committees and tenure reviewers are not likely to recommend spending precious time in this way. However, observations are absolutely essential for you to generate working hypotheses that are novel yet grounded in reality. Carve out some time to get to know your organisms. If you are too busy with classes and other responsibilities, then reserve two days before you start your experiments to observe your system with no manipulations (or preconceived notions). It often helps to do this with a lab mate or colleague. The opposite is also true; spending a whole day with no other people around and no distractions just looking at your organism can be very instructive as well. Even after you have set up your manipulations, continue to monitor the natural variation in your organisms. This will help you interpret your results and plan better experiments next year. For example, Mikaela's initial project plan for her first research project involved examining the role of fire on butterfly assemblages on forested hillsides. Poking around during her first season revealed that most butterflies were using riparian areas, a habitat that fire ecologists had largely ignored. This led to a second experiment the following year that was far more informative than the original experiment she had planned (Huntzinger 2003).

A field notebook is one tool that may help you make and use observations. It is difficult to remember all the details that you observe. Jot them down in your notebook even if they don't seem particularly relevant to the question you are addressing. Also jot down ideas that you have in the field about your study organism, other organisms it may interact with, general ideas about how ecology works, even unrelated ideas that pop into your head, even if they seem unimportant. It is amazing how valuable some of these observations can be at a later time.

An excellent way to begin a project is by observing and quantifying a pattern in nature, as we mentioned in chapter 1. Common ecological patterns include changes in a trait of interest that varies over space or time. This could be anything from a trait of individuals (e.g., beak length) to a trait of ecosystems (e.g., primary production or species diversity). Ask questions about it. How variable is the trait? Is there a real pattern to the variation over space or time? For example, are there large differences in primary production from one place to the next? What factors correlate with the variation that you observe? For example, do the differences in diversity follow a latitudinal gradient? What factors covary with this response variable (species diversity)? For example, what factors vary from the poles to the equator that could help to explain the observed pattern in diversity? It is often helpful to represent the pattern as a figure with one variable on the x axis and the other on the y axis. This representation allows you to get a sense of the pattern—how strong it is and whether the relationship between the two variables appears to be linear. At this point, an experiment can help to determine if the two variables are causally linked. If the relationship appears linear, then an experiment with two levels of the independent (predictor) variable may be appropriate. For example, if the relationship between the number of pollinators and seed set is linear, then an experiment with and without pollinators may be informative. If the relationship is nonlinear (let's say hump-shaped), then an experiment with only two levels (with and without pollinators) will not be as informative as an experiment with many levels of pollinators.

Observations are critical for meaningful experiments; in some cases they even replace experiments as the best way to gain ecological understanding. This is due in part to the unhappy fact that many processes are difficult to manipulate experimentally. Manipulative experiments must often be conducted on small plots and over short periods of time (Diamond 1986). However, important ecological processes often occur at scales that are large or have little replication. These processes may involve organisms that cannot be manipulated for ethical reasons. Other processes are simply hard to manipulate in any realistic way. For example, manipulations involving vertebrate predators are difficult to achieve with any realism. Their home ranges are often larger than the plots available to investigators. Removing predators is often more feasible than adding them, but may be unethical. Observations are often possible in these and other situations when experiments are problematic. Observational experiments still require replication and controls to be most informative (see "Manipulative Experiments" below). We discuss how to analyze observed patterns in chapter 5.

Although it's tempting to extrapolate our results from small-scale experiments to more interesting and realistic processes at larger scales, it is difficult to justify doing so. One partial solution to this dilemma is to observe processes that have occurred over larger spatial and temporal scales and ask whether these observations support our data from modeling and doing experiments at small scales. Such observations are sometimes termed "natural experiments," since the investigator does not randomly assign and impose the treatments (Diamond 1986).

Long-term data sets expand the temporal scale of any experimental or observational study. If you have the opportunity to link your work to a long-term survey, it is worth considering. For example, daily surveys of amphibians have been collected at one pond, Rainbow Bay in South Carolina, since 1978. This record has been useful for understanding the causes of worldwide amphibian declines (Pechmann et al. 1991), the consequences of anthropogenic climate change (Todd et al. 2011), and other ecological issues.

Why are observations less valued than manipulative experiments? Observations can be applied to test hypotheses, but they are poor at establishing causality. This is their main limitation. For example, we can observe that two species do not co-occur as frequently as we would expect. This may suggest that the two are competing. In the early 1970s everyone was "observing competition" of this sort because it made such good theoretical sense. However, the observed lack of co-occurrence could be caused by the two species independently having different habitat preferences that have nothing to do with current competition. Observations alone do not allow the causes of the pattern to be determined, although methods are being developed to infer causal relationships from observational surveys (chapter 5). But while limited in some ways, observations provide natural history intuition at realistic scales so that important factors for manipulative experiments and modeling can be identified.

Manipulative Experiments

Manipulative experiments vary only one thing (or at most a few). The experimenter controls that variable. If the experiment has been set up properly, any responses can be attributed to the manipulation. This approach is very powerful for establishing causality. Treatments, including controls, should be assigned randomly so that they are interspersed (Hurlbert 1984). Statistical tests can then be used to evaluate the likelihood that the observed effect was caused by chance or by the manipulation. These issues will be considered in much more detail throughout this book, particularly later in this chapter and in chapter 4.

Model Building

Modeling is an attempt to generalize, to distill the cogent factors and processes that produce the behaviors, population dynamics, and community patterns that we observe. The strengths of the approach are that the results apply generally to many systems and that the model allows us to identify the workings of the important elements. Mathematical modeling forces us to be explicit about our assumptions and about the ways that we envision the factors (individuals, species, etc.) to be related. Since we often make these assumptions anyway, writing a model almost always focuses our thinking. We all use models to organize our observations, although these are usually verbal generalizations rather than mathematical equations. The act of writing an explicit model forces us to be more precise about the logical progression that produces generalizations. Models also allow us to explore the bounds of the hypothesis. In other words, under what conditions does the hypothesis break down?

Models can be general or specific; both kinds are usually constructed of mathematical statements. General models allow us to formulate the logical links between variables. Specific models involve measured parameters from actual organisms and allow us to make detailed predictions (e.g., how much harvesting can a population sustain?).

Successful models can let us develop new hypotheses about how nature works or about how to manage ecological systems. For example, theoretical models predicted that apparent competition should be common in nature (Holt 1977). In apparent competition, two species at the same trophic level appear to be competing, while in fact one species is actually causing shared predators to become more abundant, which depresses the other species (fig. 1). Partly as a result of these modeling results, Holt and others have looked for this phenomenon in nature, and it is indeed widespread (reviewed by Holt and Lawton 1994). Models have also proven useful in designing conservation and management strategies. For example, a detailed demographic model of declining loggerhead turtles indicated that populations were less sensitive to changes in mortality of eggs and hatchlings and more sensitive to changes in mortality of older individuals than conservation ecologists had previously realized (Crouse et al. 1987). This result prompted changes in efforts to protect turtle populations, which have improved their prospects for survival (Finkbeiner et al. 2011).

In addition to helping us develop new hypotheses, models can tell us where to look for the patterns in nature. For example, Darwin observed many finches with different morphologies and life histories on his visit to the Galapagos. But he didn't record which morphologies were found on which islands. He didn't see this information as relevant at the time because he had not yet generated a model of differentiation and speciation.

Models can make logical connections easier to see. Often the consequences or results are well known and very visible but the processes that caused those results are difficult

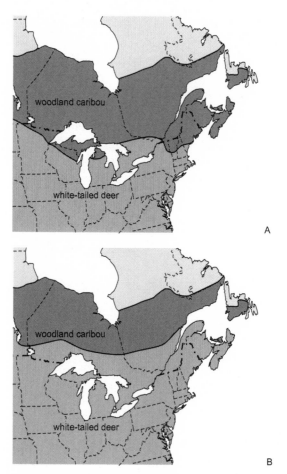

FIGURE 1. Apparent competition between white-tailed deer and caribou (Bergerud and Mercer 1989). A. Caribou historically lived in New England, Atlantic Canada, and the northern Great Lakes states (redrawn from a Wildlands League caribou range map). B. Since European colonization, records show deer expanding their range into these areas and replacing caribou (redrawn from Thomas and Gray 2002). Numerous efforts to reestablish caribou into areas where they could contact deer have failed. C. The conventional hypothesis for the declines in caribou involved competition for food

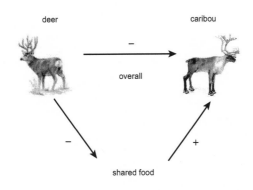

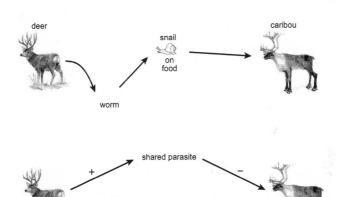

resources. More deer meant less food for caribou (shown as a negative effect of deer on shared food). This explanation has not been supported by data, and one current hypothesis involves mortality to caribou caused by a shared parasite, a meningeal worm. D. White-tailed deer are the usual host for the worm, and they are far more tolerant of infection than are moose, mule deer, and, especially, caribou. Caribou get the worms by ingesting snails and other gastropods that adhere to their food. The gastropods are an intermediate host for the worms.

to assess. Models can force us to consider alternative mechanisms when the currently favored explanation does not produce the "right result" in our modeling effort. For example, John Maron and Susan Harrison (1997) were faced with trying to explain why high densities of tussock moth caterpillars were tightly aggregated. They knew from a caging experiment that the moths could survive outside of the aggregation, although in nature the moths were restricted to the aggregation area. Spatial models suggested that very patchy distributions could arise within a homogeneous habitat if predation was strong and the dispersal distance of the moth was limited. As a result of these model predictions, they looked for predation, a previously unexamined explanation, and found that it was indeed operating.

Models of all sorts have been used to test ecological hypotheses. If a model has been constructed that includes particular ecological mechanisms, it is possible to ask whether the model fits actual data. If the fit is good, people may argue that the ecological mechanisms included in the model are probably operating. However, this reasoning uses correlation to support the implicit hypothesis that the mechanisms that provide a good fit are the ones that caused the patterns in the data. Such arguments rarely acknowledge that other mechanisms could also produce good fits to the data and may be the actual causal factors in nature. Models can be very useful as long as you don't over-interpret them.

It is unfortunate that modeling and natural history often attract different people with different skills and little appreciation for each other's approach. However, many of the most successful ecologists have been individuals who have been able to bridge these two approaches.

Why Ecologists Like Experiments So Much (Or Why We Couldn't Call This Book *The Tao of Ecology*)

The Tao is an ancient Chinese term that refers to the streamlike flow of nature. Like a stream, the Tao moves gently, seeking the path of least resistance and finding its way around, without disturbing or destroying. A Tao of Ecology might entail noninvasive and nondestructive observations of entire systems to understand who the players are and how they interact with one another. We find the Tao to be an appealing image in general and one that could be applied to ecology (see book cover). However, nothing could be farther from the approach that most ecologists currently employ. In this section, we explain why ecologists like to manipulate their systems so much.

In recent decades, ecology has come to rely on manipulative experiments. The investigator disturbs the system and observes what effect the disturbance has. This experimental approach has the advantage of providing more reliable information about cause and effect than do more passive methods of study. Understanding cause and effect is critical, powerful, and much more difficult than it sounds.

Consider the inferences that can be drawn based on observations versus those based on experiments. Observations provide us with a chance to discover many correlations; however, correlations provide limited insight into cause-and-effect relationships. One version of the old adage says

that correlation does not imply causation. Bill Shipley
(2000) points out that this is incorrect. Correlation almost
always *implies* causation, but by itself, cannot *resolve* which
of the two correlated variables might have *caused* the other.
Let us give two examples, the first from one of our life experiences and the other from the ecological literature.

The end of grad school was a time of reckoning for
Rick. The only car he had ever owned, a Chevy Vega, was
clearly falling apart, although he pretended not to notice.
His girlfriend convinced him that since he had a job lined
up on the other side of the country, and he would soon
actually have a salary, he should abandon his grad-student
lifestyle and buy another car before heading west. Red has
always been his favorite color, so naturally he was interested in a red car. However, his girlfriend had seen a graph
on the front page of *USA Today* indicating that red cars are
involved in more accidents per mile than cars of other colors. Concerned about their safety, she argued for another
color. Statistics don't lie, and red cars are more dangerous
than other cars. Her working hypothesis had the cause and
effect as "red causes danger":

Red ⎯⎯⎯⎯⎯⟶ Danger

Rick was unsuccessful in convincing her that more dangerous (sexy?) people chose red cars in the first place and
that getting a more boring color would do little to help
them:

Danger ⎯⎯⎯⎯⎯⟶ Red

In the end, Rick bought a gray car but drives a red one now (when a bicycle won't do). As this book goes to press he has luckily escaped being in any automobile accidents.

This example may seem silly, and unlikely to happen to scientists (Rick's girlfriend was a social worker). We can assure you that we have seen it repeated many times by ecologists who infer causal links from correlations. For example, Tom White made the insightful observations that outbreaks of herbivorous psyllid insects were associated with physiological stress to their host plants and these outbreaks followed unusually wet winters plus unusually dry summers (White 1969):

| unusual weather | ——— | physiological stress | ——— | psyllid outbreaks |

He argued that plant physiological stress increased the availability of limiting nitrogen to the psyllids he studied and to many other herbivores (White 1984, 2008). So essentially he hypothesized a causal connection between these correlated factors:

$$\text{weather} \longrightarrow \text{stress} \longrightarrow \text{increased N} \longrightarrow \text{herbivore outbreaks}$$

However, the actual causal links could be different. For instance:

$$\text{weather} \longrightarrow \text{herbivore outbreaks} \longrightarrow \text{stress} \longrightarrow \text{increased N}$$

Or perhaps weather influences some other factor that then causes herbivore outbreaks, without involving the host plant:

weather $\longrightarrow$ plant stress
$\searrow$ reduced predation $\longrightarrow$ herbivore outbreaks

Without manipulative experiments, it is difficult to establish which of these causal hypotheses are valid and important. However, if microenvironmental conditions, physiological stress, available nitrogen, herbivore numbers, and predator numbers can all be manipulated, it will be relatively easy to determine which of these factors cause changes in which others. In the end, White's intuition got him fairly close to the truth. A review of experimental studies suggests that herbivores, especially the sap feeders that White studied, are negatively affected by continuous drought stress, but that intermittent bouts of plant stress and recovery promote herbivore populations (Huberty and Denno 2004).

Replicated manipulative experiments have the potential to provide more definitive evidence about causality, but unfortunately many ecological problems are not amenable to experimentation. New techniques are being developed that can provide inference about causal relationships from observational data (Shipley 2000, Grace 2006). These techniques involve directed graphs (the diagrams with arrows shown above). Once we have specified a causal path or directed graph, we can predict which pairs of variables will be correlated and which pairs will be independent of one another. These techniques allow us to build models that estimate the probability of causation from correlations in the data. We can then discard causal models that don't fit our observations.

These methods are not difficult to use, but are much less well known than inferential statistics such as analysis of variance (see chapter 4). This correlational approach is most useful when one model matches the observed patterns more accurately than alternative models, which is often not the case in ecology. The jury is still out on how often and severely the assumptions will limit the applicability of these new methods. We will return to directed graphs in chapter 5.

In summary, ecologists love manipulative experiments because we love understanding causality. Regardless of your philosophical persuasion on this issue, the truth remains that it is easier to publish experimental work than work relying on observations and correlations. Observations can address patterns at larger scales than experiments, but if you have a choice between observations and experiments to ask the same question, experiments are more powerful and convincing. However, not all experiments are created equal. Experiments are only as good as the intuition that stimulated the experimenter to manipulate the few factors that he or she has chosen. Experiments are limited by this initial intuition and by problems of scale and realism. In addition, natural history observations can provide the intuition to design meaningful experiments and provide information over larger areas and longer time frames than an experimenter can handle with manipulations. Another approach, modeling, can provide generality, suggest results when experiments are impossible, project into the future, and stimulate testable predictions.

Whenever possible, you should integrate several of these approaches to pose and answer ecological questions. One approach can make up for the weaknesses of another. The best of modern ecology combines observations, models, and manipulative experiments to arrive at more complete explanations than any single approach could provide. You are after the best cohesive story you can put together.

Using Experiments
to Test Hypotheses

As we discussed in chapter 2, manipulative experiments are valuable in ecology because they can help to establish causality. The experimenter manipulates the treatments and observes the effects of the manipulations. In the next chapters, we consider experimental and statistical techniques that are used to evaluate cause-and-effect relationships in ecology.

Experimental Requisites

Unambiguous interpretation of causality is dependent on several requirements: (1) appropriate controls, (2) meaningful treatments, (3) replication of independent units, and (4) randomization and interspersion of treatments (Hurlbert 1984).

Controls

Controls are treatments against which manipulations are compared. Biological systems change over time. As a result, we cannot simply compare our experimental units before and after we apply treatments. Any differences that we observe before and after application of treatments could be caused by the treatments but also by other changes that occurred during this time period. Since both the controls and the manipulated treatments experience whatever changes occur over time, the effects caused by treatments can be separated from those caused by other changes. For example, we could experimentally add french fries (or environmental estrogens) to the diets of male deer during fall and winter. We would observe that they shed their antlers in March. Without controls we might conclude that the added experimental french fries had caused the antlers to be shed. However, controls without added fries would help us recognize that other seasonal changes were responsible for the antler shedding that we had observed. In this case, our control treatment is the population of male deer with no added food, essentially "no treatment."

The logical control in some experiments is not necessarily "no treatment." For example, if we wanted to evaluate effects of elevated CO_2 on plant growth, we might want to compare plants grown in an environment with CO_2 levels that are projected for 2050 with controls grown under current ambient conditions (no treatment). However, a more meaningful control might involve growing plants under conditions with 25% less CO_2 than current ambient conditions, since this is the estimated level before the Industrial Revolution.

Meaningful Treatments

When we impose experimental treatments, we often change things other than the factors we are hoping to manipulate. For example, if we want to evaluate the effects of herbivores on plant traits, we could set up the following replicated and randomly assigned experimental treatments: (1) plants caged with herbivores and (2) control plants that lack cages and herbivores. This experiment seems straightforward and easy to interpret. However, any differences that we observe between these two treatments could be caused by the presence or absence of herbivores or could be caused by the cages themselves. The cages may alter the microenvironment experienced by the plants; exclude beneficial organisms (pollinators) from the plants; eliminate harmful organisms (plant pathogens or other herbivores) from the plants; interfere with the normal behavior of the herbivores so that their effect is greater or less than it would be without a cage; or interfere with the normal behavior of the plants so that their usual developmental or reproductive schedules are disrupted. And on and on. The message is that establishing cause and effect by a simple manipulative experiment can be surprisingly difficult.

There are several ways to minimize the artifacts of the cage example. You should design a cage treatment that causes as few as possible of these unwanted secondary effects. In addition, the secondary effects you can't avoid can be tested to evaluate their likelihood as confounding artifacts, such as by including additional treatments that control for these potential artifacts. For example, cages can be constructed of different mesh sizes that allow some of the smaller organisms to come and go freely but block larger ones. Alternatively, cages can sometimes be left open at the top or bottom so that some of the microenvironmental

side effects caused by the cage are included in the controls. It is often a good idea to attempt to impose the treatment in several different ways. Each of the different im-
positions may have its own side effects. So, for instance,
another way to exclude small herbivores is to treat plants
with selective pesticides. Such a treatment is likely to cause
its own artifacts. However, the artifacts associated with pesticides are probably different from those caused by caging.
If you find that herbivores have a consistent effect on
plants regardless of how they are experimentally manipulated, you can feel more confident that your conclusions
are real and robust. You can often avoid erroneously assum-
ing that your treatments have caused your effects by thinking first about potential side effects of your treatments and
attempting to include them in controls.

  Care should be taken in deciding on appropriate treatments and controls. Biological processes may or may not
be easily mimicked by manipulative treatments. Consider
the example of an experimental treatment that attempts to
mimic the effects of fire on plants. Some ecologists mimic
landscape-scale fires with small-scale experiments by conducting fires in 1-m^2 fireproof arenas, placed around vegetation. These experimental fires only get a fraction as hot
as real fires, combust only a small proportion of above-
ground biomass, are generally conducted in a different
season than wildfires, and so on. Similarly, the easiest and
most straightforward way to mimic the foliage-chewing behavior of herbivores is to cut leaves with scissors. For some
plants, clipping with scissors adequately captures the effects of herbivores. However, how the leaf area is removed,

in one big bite or many small ones, whether the veins are severed, and so on, can greatly influence the effect of clipping on the responses of plants (Baldwin 1988). In addition, components of saliva from particular herbivores have been found to have profound effects on some host plants (Felton and Eichenseer 1999). Carefully designed treatments involving actual organisms and actual processes, when feasible, are preferable to more artificial treatments.

When designing experimental treatments, attempt to span the natural range of variation. For instance, suppose we want to evaluate the effects of herbivores on plants. Repeatedly defoliating the plants may produce dramatic effects; however, this result will tell us little about the effects of real herbivores on plants if repeated defoliation does not occur in nature. Conversely, picking only a single, modest level of damage may cause us to underestimate the effects of real herbivory. The best bet here might be to use a range of treatment levels that spans the range of damage that is naturally encountered. We can sometimes gain interesting insights by including treatments outside of the current range that represent projections of future conditions.

Common sense and preliminary observations are often better yardsticks for meaningful treatments than guidelines that can be found in the literature. Nonetheless, you can get some useful ideas about methods in the following references: Elzinga et al. (2001) for general sampling and data management; Sutherland (2006) for general sampling with more detailed information about specific taxa; Moore and Chapman (1986) for sampling plants; Kearns and Inouye (1993) for methods used in studying pollination and plant fitness; Southwood and Henderson (2000) for sampling insects; and Wilson et al. (1996) for methods used to sample mammals.

Replication

It is important to replicate independent experimental units of each treatment and control so that you can separate the effects of the treatment from background noise. Imagine for a moment that there is only one replicate (independent sample) of each treatment. It will be impossible to determine whether any differences between the treatments (manipulations) are really due to differences caused by the experimental manipulations, rather than by other differences between the particular individuals sampled for each treatment. With only one independent replicate, no amount of subsampling or precision will help establish causality because the factors that affect one subsample may also affect others. However, if many independent replicates show a difference between treatments, we can be more confident that this difference was caused by the treatments. For example, Dave Reznick and John Endler (1982) observed that guppies from a Trinidadian stream that had high risk of predation from larger fish had different life histories than guppies from a stream with low predation risk. Fish with high risk of predation became adults more quickly and devoted more resources to reproduction than those with low risk of predation. Collecting many guppies (subsamples) from one site of each treatment did little to improve the inference that predation was the cause of the life history differences. Instead, Reznick increased replication in both an observational survey and a manipulative experiment. First, he found many different streams in Trinidad and characterized the life history traits of the guppies as well as their risk of predation. This gave him more confidence that the relationship between predation and guppy life histories was a real one. Second, he moved guppies from streams with high predation to streams with low predation (Reznick et al. 1990). Descendants of the transplanted guppies had traits that matched those of the guppies that had lived in low-predation streams for generations. Furthermore, this experiment was repeated in streams in two different river systems. This work is convincing because the results were consistent across many replicates.

Getting independent replicates is not always easy. In practice, try to separate independent replicates by enough space so that conditions for one replicate do not influence conditions for another replicate. This distance will be determined by the organisms in question—generally larger and more mobile organisms will need greater spacing than smaller, more sedentary ones.

The point to remember is that, in general, the more independent replicates you have, the greater your power is to detect treatment effects.

Because of time and effort constraints, replication often comes at the expense of precision of each estimate. This is fine. It is almost always better to take many samples, each with little precision, than to spend your limited time making sure that any given sample is accurate. The central limit theorem of probability can help you out here. If you take a large number of unbiased estimates, each one very imprecise, you will quickly arrive at an estimated mean value that is close to the actual mean value. This is an impressive party trick. Get your friends to estimate the size of some object, say a window. Individual estimates are likely to be far from the actual value (man, do some people have bad judgment). However, the mean from a group of about 30 partyers will be astonishingly close to the actual value. The message from this exercise is clear. Always go for as large a sample size as you can get, even if each of your samples is sloppy and noisy. A large, unbiased set of samples will average over the noise and bail you out. This advice is supported by analyses of real ecological data (Zschokke and Ludin 2001). Imprecise measurement had surprisingly little effect on ecological results, suggesting that limited time and resources are much better invested in more replicates than in greater precision of measurements.

A large sample size can rescue imprecise measurements, but it cannot rescue biased measurements. For example, imagine asking a group of 30 partyers for their estimates of the age of the earth. The mean value will be quite different if the partyers are geologists or religious fundamentalists. Increasing the sample size will not alter this bias.

Students who are just starting to do research often want to know how large a sample size they will need. There is no easy answer to this question; it depends on the size of the difference that you are interested in detecting and how much noise there is. Yes, but let's get real: that knowledge doesn't help you determine how large a sample size to shoot for. As a general rule of thumb with no other information, we always try to get 30 independent replicates of each treatment. If 30 is impossible, we may be able to get by with 15. Below 15, we start to get anxious.

Some experiments do not lend themselves to lots of replication. For example, conservation questions at the landscape scale can be replicated only a few times. We have conducted experiments in large plots that excluded different mammalian herbivores (Young et al. 1998). In these cases, each treatment was replicated only three times. Here subsampling allowed us to get a more precise estimate of the response in each plot so that statistically significant differences were revealed (Huntzinger et al. 2004). It helped that the effects we were looking for were relatively huge; no amount of subsampling would have made these low levels of replication work for small effect sizes. Sometimes it is impossible to replicate your treatments at all. In these cases, traditional interpretations of statistical tests are probably inappropriate (although this suggestion is contentious— see Oksanen 2001). Results without statistical tests are difficult to publish by themselves but may accompany smaller-scale, replicated studies to provide biological realism to those statistically significant results.

On the subject of sample size, statisticians always recommend that you collect preliminary data to determine the appropriate sample size. They argue that doing this will save you time in the long run. Although this is sound advice, we have never heeded it. It probably makes more sense for lab scientists than field scientists and for those blessed with patience than for the three of us.

Having a large number of replicates increases your power to detect differences caused by your treatments. However, high replication comes at the expense of the size of each replicate. In other words, if you want to have many replicates, each of those replicates is going to be small. This can be a serious problem because some processes operate only at particular spatial scales. For example, you are likely to get one result placing a predator in a large arena where it can set up a normal territory and its prey can replenish themselves, and another result in a small arena where it will behave abnormally and quickly eliminate its prey.

Despite small replicate sizes, high levels of replication have advantages. Having replicates not only gives you statistical power but also makes common sense. With only one or a few replicates of each treatment, you have no idea if the differences you observe were caused by your treatments or by some other factor. Independent replicates that are not biased with respect to the treatments allow you to avoid spurious interpretations.

However, insufficient replication is just one possible source of an incorrect interpretation. Experiments conducted using spatial and temporal scales that are too small can also lead to incorrect inference. Since replication almost always comes at the cost of scale, some ecologists argue that our field has leaned much too far in the direction of replication and that scale should take priority (Oksanen 2001). As previously mentioned, we often try to deal with this problem by working at two spatial scales—conducting a highly replicated manipulative experiment with small units and a poorly replicated experiment or observation with large units. If the answers are similar at both scales, the conclusion is much stronger. The problem of determining the scale at which you should work is particularly acute when dealing with resource managers, growers, or agriculturists, who are not very hung up with statistical tests but won't listen to results conducted on small plots. Most academics will not listen to results (or publish them) if they are not statistically significant, and higher replication makes statistical significance more likely. The solution? You can't please these two audiences simultaneously. Conduct two different experiments, one with high replication and the other with a large spatial scale. It probably doesn't matter which scale you start with; there are unique advantages to both.

One way to visualize the scope of your experiments and observations is to plot them on a graph with spatial scale as one axis and temporal scale as the other axis. In this way you can clearly see the spatial and temporal range that your manipulative experiments, natural experiments, and models cover. For example, Schneider et al. (1997) wanted to understand the multiannual population dynamics of a bivalve mollusk at the scale of an entire harbor (368 km^2). Their experimental units were 13 cm cores taken over a 30-second period. These units were repeated over an area of about half a kilometer^2 during a 28-month period, which allowed them to greatly expand the scope of their experiment. Nonetheless, making inferences about dynamics that occur in the harbor on the scale of decades required considerable extrapolation (fig. 2). They combined this experiment with a model of the entire harbor. They used information from the small-scale experiment to suggest model parameters and information from the model

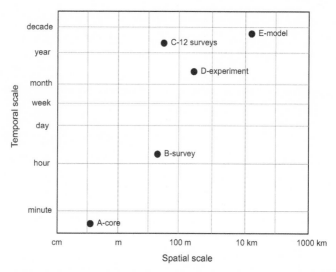

FIGURE 2. A scope diagram showing observational, experimental, and modeling approaches used to characterize changes that occur to harbors over decades (Schneider et al. 1997). The diagram makes explicit the spatial and temporal extent of each observation or experiment, as well as the scale over which Schneider et al. would like to extrapolate. The observational and experimental unit for all sampling was a 13-cm-diameter core taken for 30 seconds (A). A survey included 36 cores at six sites (B), repeated during 12 monitoring visits (C). An experiment was conducted in which ten cores were taken at each of two experimental plots, each of which was visited nine times over a 100-day period (D). Schneider et al. also constructed a model at the scale of the entire harbor (E) and attempted to extrapolate downward to compare their model results with those from observations (surveys) and experiments.

to help interpret their experimental results and suggest further experiments.

Since manipulative experiments in ecology today are almost always conducted at spatial and temporal scales smaller than our ideal, it is worth considering what effect this has on our worldview. Small-scale experiments have led us as a group to believe in local determinism, that is, that the processes we can manipulate on a small scale mold the patterns that occur at larger scales (Ricklefs and Schluter 1993). However, this expectation is likely to be simplistic when we look at real communities. For example, local processes such as competition and predation tend to reduce species diversity, whereas larger-scale regional processes tend to increase diversity through movement and speciation. One way for you to see that small-scale experiments don't capture all of the important processes is to place a barrier around a local area and observe whether all of the species persist. In most cases, they don't. Even the largest parks, such as Yellowstone and the Serengeti, are too small to maintain a full complement of species over the long term (Newmark 1995, 1996).

Achieving temporal replication is also challenging. It is difficult to replicate experiments over more than a few years. Ideally we would, because conditions (weather, species abundances, etc.) vary enormously from year to year. Yearly variation has been found to have strong effects on the results of ecological experiments, but temporal replication is rarely addressed in the design or interpretation of experiments (Vaughn and Young 2010).

In summary, an appreciation for larger-scale processes such as the interactions with other organisms in the species' ranges and interactions that occur over longer time frames can help our thinking, although these processes are difficult to study experimentally (Ricklefs and Schluter 1993, Thompson 1999).

Independent Replicates, Randomization,
and Interspersion of Treatments

Replication serves a useful purpose in experimental design only if the replicates are spaced correctly. For instance, if all of your high-nitrogen replicates also happen to be in a swampy area and your controls are in a drier upland area, then you might conclude that nitrogen caused effects that were actually caused by the swamp. Therefore, experimental replicates must be independent of one another (Hurlbert 1984). Independent replicates make it likely that the noise associated with each treatment is unbiased and that the treatments are on average similar in all ways except for the treatment effect. Consider a common design for experiments that examine the effects of biological control agents (insect predators) on greenhouse pests. The greenhouse is physically divided in half with a screen barrier down the middle (fig. 3A). Predators of the pest are released into one half of the greenhouse, containing, let's say, nine plants. No predators are released into the control half of the greenhouse, also containing nine plants. It is incorrect to assume that each of the two treatments is replicated nine times. If one side of the greenhouse is different from the other, then all of the plants of each treatment will experience those differences. In essence, there is only one independent replicate of each treatment.

The way to get independent replicates that are unbiased is to intersperse the replicates of the various treatments. In other words, placement of the two treatments must be all mixed up (fig. 3B). If one side of the greenhouse is sunnier or windier than the other, these differences will not

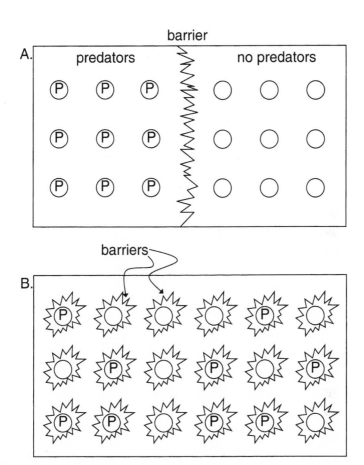

FIGURE 3. Experimental designs to evaluate the effects of predators on greenhouse pests. A. A pseudo-replicated design with one barrier separating the two treatments and only one independent replicate of each treatment. Treatments are not interspersed. B. A proper design with 18 barriers and 9 independent replicates of each treatment.

be confounded with treatment differences. Any observed differences associated with the treatments will probably be caused by the treatments. Of course, setting up this design is going to require many more screen barriers than the design that divides the greenhouse in half. In summary, by assigning treatments randomly, we can differentiate between the possibility that our treatments caused the results that we observed and the possibility that some other factor caused the results.

The best way to assign treatments is by using a random number generator. This is often most conveniently done at home before going out to the field. Free random number generators are available on the Web. If you don't have access to the Internet when you're in the field, a deck of cards or a telephone directory can provide random numbers (use the last few digits; the first few are not random). For two treatments, use red/black (cards) or even/odd (phone numbers); for three treatments, use three suits or only the phone numbers that end in 1, 2, and 3. In any of these scenarios, make sure that each treatment is assigned the same number of replicates.

Random assignment does not mean assigning every other individual to each treatment the way you deal cards, nor does it mean going along and haphazardly assigning treatments. Both of these methods are acceptable, but they should be identified as alternating (regular) or haphazard assignment of treatments, and both will reduce the power of your statistical inference. Randomization is usually an effective way to achieve interspersion of treatments.

If your random assignment of treatments fails to provide treatments that are well interspersed and matched for other factors, the experimental units should probably be reassigned (Hurlbert 1984). In other words, if by chance many of one treatment wind up being on one side of the plot (which turns out to be drier) and many of the second treatment wind up being on the other side of the plot (which turns out the be wetter), moisture and treatment will be confounded. It is worth assigning treatments randomly a second time even if you don't suspect that the two sides differ in moisture or in some other unmeasured variable, since there are so many confounding spatial factors that you might not anticipate.

If you know that there is an environmental gradient in your study plot before you assign treatments, it is a good idea to block your experimental site and assign treatments within blocks (Potvin 1993). For example, if you know that one side of your plot is on a slope and the other side is on level ground, divide (block) the field into two subplots (slope and level) and then randomly assign an equal number of replicates of each treatment to each of these two blocks. Blocking can reduce the noise caused by the environment and give you more power to detect an effect of your treatments than can a completely randomized design in which there are large effects of blocks. However, there is a cost to blocking. Blocks decrease the error degrees of freedom (statistical power) in your analysis; the more blocks, the larger this effect. If blocking does not accurately match the environmental heterogeneity that is important to the outcome of your experiment, then blocking decreases power relative to a completely randomized design.

Many ecological studies are designed with inadequate replication and interspersion or other flaws in design, analysis, and interpretation. If we identify such a problem, it is a natural temptation to disregard or trash the results, a phenomenon our friend Truman Young calls "pseudo-rigor." We suggest that this temptation should be held in check. A study that has a design flaw may be subject to alternate explanations; on the other hand, the conclusions may be correct, and the study almost certainly has some biological intuition to offer. Too often people don't analyze data that they have collected because they feel their design wasn't perfect. This may be true, but the work almost certainly has something to teach you (and others). Let's keep things in perspective. Even the best experimental study is subject to alternative explanations because only a limited number of factors were considered. In chapter 2, we discussed Tom White's interpretation of the causal relationships among weather, plant stress, and outbreaks of psyllids. Since he only had a correlation, he couldn't establish cause and effect. Nevertheless, it would be a mistake to disregard his insights. Based on the information he had, he couldn't be sure of causality, but experiments done in the interim suggest that he was certainly on the right track (Huberty and Denno 2004). Without long-term data and biologists who have an intuitive grasp of their organisms, we won't do the proper experiments, with or without statistical "rigor."

Lab, Greenhouse, or Field?

Ecologists are often tempted to work in environments that minimize unwanted variance or "noise." The reasoning is that controlled environments enable us to vary single factors in order to isolate their effects (Potvin 1993). Ecologists do this to varying degrees and in different ways. We choose field sites that are well matched and as similar to one another as possible. We move into the greenhouse, where abiotic conditions can be controlled and made similar for all replicates. Sometimes we conduct experiments in small growth chambers, aquariums, and lab "microcosms," which provide even more environmental control. The real world (field) can be so complicated that it can be difficult or impossible to perceive patterns because of all the noise. A simplified, controlled environment can reduce this noise and allow us to see the signal, test predictions, or get at mechanisms that would not be possible in the field. In addition, working in these controlled environments is often more convenient than working in the field. Controlled environments, such as greenhouses or growth chambers, are often close to where we have other obligations (e.g., our classes or families) and allow us to conduct experiments during times when natural systems may be inactive. They may also be close to the equipment we need to impose treatments or measure responses.

This control and convenience come at a large and often unrecognized cost. First, controlled environments are generally far more variable than we imagine (Potvin 1993). In our experience, plants and insects grow very differently on one side of a greenhouse bench than on the other. This variance in the greenhouse is often larger than we have encountered in the field.

Second, working under controlled conditions is unrealistic. For example, plants in the greenhouse or organisms in aquariums routinely experience outbreaks of pests that remain at quite low levels under field conditions. In addition, anything that you learn in the field is potentially interesting and important because it happened in the real world. It can lead to new research directions. For example, at the end of his thesis research, Rick set out to learn about mortality factors that could affect populations of cicadas in the field and noticed induced plant responses that killed cicada eggs (Karban 1983). This unexpected turn of events stimulated the questions about induced resistance that he asked for the next 30-plus years. Years later, working in the field instead of the greenhouse paid off again for Rick. While he was asking questions in the field about induced resistance in wild tobacco plants, an unexpected frost damaged many of these plants. This seemed like a disaster at first, but he learned that induced plants are more susceptible to frost and that this risk may represent an unappreciated cost of induction (Karban and Maron 2001). In contrast, when the temperature controls malfunction in his growth chamber, he doesn't learn anything useful about plant responses to real variation in temperature.

Being opportunistic is likely to pay off when you are working in natural conditions. After conducting many repetitions of a lab experiment, Rick found that the strength of induced resistance varied from one experiment to the next. To his disappointment, he figured out that some of this variation was due to using different pot sizes (Karban 1987). Plants in pots dry down at different rates, and pot-bound plants are less inducible. Who cares? This result provides very little inference about how organisms work in the real world.

Often ecological phenomena do not transfer from the field to the lab. For example, Henry Horn studies the mosses and lichens that grow on boulders near Princeton, New Jersey (personal communication). These small organisms survive droughts because the surface temperature of the large boulders lags behind the air on a daily cycle, which allows them to extract water from the air in the early morning. This process does not occur in the lab or greenhouse, and this system cannot be studied indoors in a realistic manner.

Laboratory experiments are usually conducted under conditions that are simplified and controlled by intent. Even if you are able to set up the experiment and answer the question that you posed, you cannot know how well it depicts similar processes in nature. The solution to this problem is to link lab and field studies. They each can provide unique information but also have unique limitations. Field observations and experiments should be followed by lab studies to learn more about the ecological mechanisms that could cause the field result. In turn, lab studies should be followed by field studies and "natural experiments" to learn whether the lab results are realistic and whether they hold at larger spatial and temporal scales (Diamond 1986).

CHAPTER 4

Analyzing Experimental Data

Hypothesis Testing and Statistics

The first step in doing research is to have a clear question or hypothesis in your mind. If you just have a vague interest in a system, an organism, or an interaction, you are not ready to do experiments; however, this would be a great time to poke around so that questions start to form in your mind. You must be able to formulate your ideas into a clear question. Without a clear question, there is no end to the data (relevant or otherwise) that you may feel compelled to collect. Good research is a bit of a balancing act: developing and pursuing a clear and focused question while keeping your eyes open for unexpected answers and new ways to rephrase the question.

A clear question stimulates relevant experimental manipulations and statistical analyses, rather than being the result of them. Observations of natural patterns are a good way to generate questions. For example, you might observe pandas and bamboo longhorned beetles interacting, and suspect that species A (pandas) reduces the population size of species B (beetles). Your testable working hypothesis is that the population of beetles will be lower in the presence of pandas than in the absence of pandas. Because you have a clear hypothesis in mind, you are now ready to design an experiment. You can conduct a manipulation, removing pandas from half of your plots and keeping the other half as unmanipulated controls. You can measure populations of beetles in these two treatments and compare the difference that you measure against the difference that would be expected by chance. Based upon a statistical analysis, you can determine the likelihood that your null hypothesis that pandas don't reduce beetle populations can be rejected. In short, the question should drive the experimental design and statistical analysis.

Statistics allow you to evaluate whether the differences caused by your treatments are likely to be real differences or are likely to reflect random noise. While you may run across older studies without proper controls, or replication, or even statistics, it is virtually impossible to get experimental studies published nowadays without statistical analyses.

We often confuse statistical significance (indicated by a probability level) with biological significance (indicated by the size of the effect). For example, we might hypothesize that diet affects the body size of rodents. If we feed two groups different diets, we can be fairly certain that the two will not grow to the same size. What we really want to know is whether they grow to sizes that are different enough to consistently observe and to be worth caring about. Very small differences can be statistically significant but not produce consequences that are ecologically important. Remember that your real interest is in biological significance. In other words, we want to know if effects are "biologically significant," but we often use "statistically significant" as a proxy.

We consider both statistical significance (represented by p-values) and biological significance (represented by effect sizes) when we evaluate and present our experiments (see chapter 8). It is insufficient to report that two populations were different and give a probability value ($p < 0.05$ or $p = 0.023$) or to report that differences were not significant (ns). Along with any significance test, report the effect size. This can be done by showing a picture (perhaps a bar graph with the means and standard errors for each of the populations), or by reporting, for example, that ants were 35% more numerous when elephants were absent. Box 4 shows how to calculate effect size. When interpreting your own results and those in the literature, separate statistical significance from biological significance. There is no hard and fast rule for interpreting biological significance. A doubling of the number of mountain lions in a particular area may have a large biological effect on the deer that live there, but a doubling of the number of mountain lions may make little biological difference to the direct nutrient inputs as carcasses when they die.

Although statistical analyses are absolutely essential for ecology to progress as a science, the emphasis in ecology on significance tests may be a bit overly zealous (Yoccuz 1991). By a convention of the discipline, we have decided to consider two populations to be different if the probability of their being the same is less than 0.05. Whether we find this magical 0.05 threshold depends on the variability of the trait within the populations and on our sample size.

Box 4. *How to calculate effect size*

The effect size is a measure of the magnitude of experimental effects. Consider a simple experiment in which large herbivorous mammals are excluded from some large-scale plots but natural densities of herbivorous mammals are allowed in others (the controls). Such an experiment could tell us, for example, about the consequences of local extinctions of large mammals (such as zebras) on populations of grasshoppers in Kenya (Huntzinger and Augustine, unpublished data). Grasshoppers eat mostly the same food resources as the mammalian herbivores in this African savanna, and the grasshoppers were more abundant in plots without mammals than in those with mammalian herbivores (see figure and table). We can report that this difference is statistically significant ($p = 0.014$), although the p-value by itself doesn't tell us anything about the size of the effect that mammals had on grasshoppers. One way to describe the size of the effect is to report the absolute difference in grasshoppers between the two treatments:

$$\left| \text{mean}_{exclosures} - \text{mean}_{controls} \right| = \text{absolute difference}$$

In this case, the absolute difference in mean grasshopper numbers per sample is $\left| 15.00 - 6.97 \right| = 8.03$. This absolute difference between treatments is not as useful as comparing the change in means caused by the treatment relative to the control:

$$\frac{\left| \text{mean}_{exclosures} - \text{mean}_{controls} \right|}{\text{mean}_{controls}} = \frac{\text{relative}}{\text{difference}}$$

In this case, removing mammalian herbivores increased grasshopper numbers by 115%. This gives us the sense

Box 4. Continued

that mammalian herbivores have a huge impact on grasshopper abundance. (We could conceivably report that controls with mammalian herbivores had 54% fewer grasshoppers than experimental plots, though this generally makes less sense than reporting differences relative to the unmanipulated controls as we did in the first case.) A third useful technique is to report standardized effect sizes scaled by a measure of the variance or noise involved with measuring them. This is often done by calculating the difference between the means divided by their pooled standard deviation (data in the table):

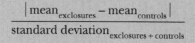

$$\frac{\left| \text{mean}_{\text{exclosures}} - \text{mean}_{\text{controls}} \right|}{\text{standard deviation}_{\text{exclosures + controls}}}$$

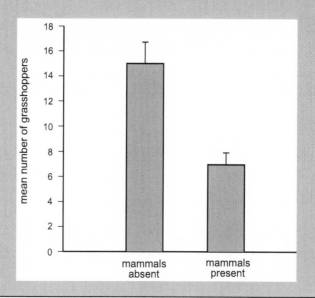

For grasshoppers, this would be $(15.00 - 6.97) / 4.89 = 1.64$. Such effect sizes are unit-less, which allows us to compare the effects found in different studies and using different response variables. Again, we see that this is a relatively huge effect.

Experimental Parameter	Value
Exclosure (no mammals)	
mean number of grasshoppers	15.00
standard error	1.71
sample size	3
Control (with mammals)	
mean number of grasshoppers	6.97
standard error	0.94
sample size	3
Both treatments (exclosure & control)	
mean number of grasshoppers	10.99
pooled standard deviation	4.89
sample size	6

If we have a very large sample size, two populations can be statistically different with means that are quite close. On the other hand, if our sample size is small, two populations that are actually quite different will not appear significantly different at the 0.05 level. It's baffling that a group of intelligent and thoughtful people can enslave themselves to this essentially arbitrary number.

We know that no two populations are identical, just as no two people are. When we test a null hypothesis that two populations are the same, we are not calculating the probability that they are truly identical but rather the probability that we can detect a difference between them. Statistical significance is really a property of the organisms, the data, and the experimenter's ability to make distinctions. Unfortunately, the 0.05 threshold has become an absolute wall between "real" results and "negative" results. Why should we be allowed to say that two populations are truly different if $p = 0.049$ but not be allowed to say much of anything if $p = 0.051$? In both cases our inference about the populations being different will be wrong approximately 5 times out of 100 (see below). You should be aware of the arbitrary nature of the 0.05 threshold and interpret results accordingly. Whenever possible, give the calculated p-value rather than reporting that p is greater or less than 0.05. If $p = 0.001$, you can be more confident that the result was not caused by chance than if $p = 0.05$. Similarly, your confidence about $p = 0.06$ should be different from your confidence about $p = 0.60$.

Even when we use statistical inference correctly, we sometimes come to the wrong conclusions. For example, Matthews (2000) found a strong positive relationship between human birth rates and the number of breeding pairs of storks in 17 European countries (fig. 4). The cause-and-effect relationship between storks and human birth rate is unclear. Including other factors in the analysis, such as the current human population or the area of the country, makes the relationship between storks and birth rate disappear. However, the hypothesis that storks bring human babies gains more support from surveys conducted in Ber-

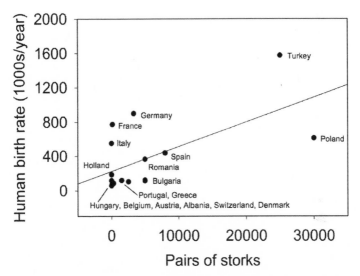

FIGURE 4. The relationship between the number of stork pairs and the human birth rate in 17 European countries, 1980–90 (redrawn from Matthews 2000). Countries with more stork pairs had higher birth rates (Births in 1000s = 0.029 Stork pairs + 225, r = 0.62, df = 15, p = 0.008).

lin, Germany, between 1990 and 1999 (Hofer et al. 2004). There was a strong positive relationship between the number of pairs of storks recorded in Brandenburg (the countryside surrounding Berlin) in each year and the number of non-hospital human births in Berlin during that year ($r^2 = 0.49$, n = 10, p = 0.024). No such relationship existed between the number of hospital births and stork pairs ($r^2 = 0.12$, n = 10, p = 0.32), presumably because storks don't bring babies to hospitals. The conclusion that storks bring babies is an example of a type I statistical error: concluding that a relationship exists when, in fact, it doesn't. Ecologists are likely to make type I errors five times out of a hundred if we accept a threshold of p = 0.05. Type II errors occur when we fail to conclude that a factor is significant when, in fact, it actually is. For example, we might conclude that sexual intercourse has no significant relationship to having babies because there are many cases in which intercourse does not result in a baby. Type II errors are probably much more common in ecology (and family planning) than type I errors.

Alternative Hypotheses

Scientists have been urged to design null hypotheses that can be critically tested and rejected (Popper 1959, Platt 1964). This is sound advice, though it doesn't work very well for many ecological questions (Quinn and Dunham 1983). Many ecological hypotheses are not simply true or false. For example, suppose we are interested in asking what the role of competition is in structuring communities. This is not a question that can be falsified by conducting a simple experiment. Evidence for competition can be found in most systems. However, so can other processes like predation and parasitism, disturbance, and so on. Instead of rejecting hypotheses and asking yes/no questions, we should be weighing alternatives and asking, for instance, how important competition is relative to other processes. We are more interested in discovering the size of the effect caused by competition and comparing it to other drivers than trying to reject a null hypothesis that it does or does not operate. In ecology, unlike some other scientific disciplines, the principles are not universal. Finding a single counterexample would make us rethink our working hypothesis about the force of gravity. However, finding a single counterexample does not disprove our ideas about competition. Similarly, ecological hypotheses are rarely mutually exclusive. Even if competition is found to be important, predation may also be important.

Too often ecologists have become advocates of a polarized point of view (e.g., density dependence versus density independence) and then spent their careers tenaciously defending it. Ecologists should strive to explicitly develop alternative hypotheses to explain the patterns they observe (Platt 1964). By developing a set of alternatives, you can avoid becoming emotionally attached to the hypothesis that you selected initially. Rick's son Jesse built a crude leprechaun trap in kindergarten, baited it with "gold" rocks, and placed it outside on the eve of St. Patrick's Day. The next morning, as Jesse checked the trap, one of the rocks dropped into the grass, an event unobserved by him. He was immensely excited by this evidence that the leprechaun had visited, taken one of the pieces of gold, and gotten away. The next year, he built a more sophisticated trap. Although the leprechaun failed to visit in this second year, he still felt confident of their existence based on the previous season's results. It is unfortunate that Jesse did not consider alternative hypotheses for the rock's disappearance. It's worth wondering how many leprechauns we each catch in our research careers.

We are all under pressure to produce significant results. Testing alternative hypotheses is an efficient way to ensure that you have something to say when you get done. If you become enamored with a particular process and conduct experiments focused only on that process and its potential mechanisms, you'll have a story to tell only if that process turns out to be as important as you thought it was.

If you start with a list of alternative hypotheses, you can reduce this stress because you are much more likely to turn up something interesting. Often after you have tested a hypothetical ecological mechanism, it becomes clear that there are alternative mechanisms that could also have contributed. Our friend Kevin Rice recommends avoiding the "house of cards" research program. If all of the secondary hypotheses that you are interested in testing require a particular outcome to be true in your initial hypothesis, then you put yourself under too much pressure to demonstrate your primary hypothesis, whether or not it is actually real. If, instead, you consider alternative hypotheses, you have something to talk about no matter what you find. Box 5 provides suggestions for generating alternative hypotheses in ecology.

Answering yes/no questions will often take the form of rejecting hypotheses, but many ecological hypotheses cannot be rejected in this way. Instead, we have argued that it may be more useful to devise a list of alternative hypotheses, acknowledge that most or all of these working hypotheses may be valid, and then attempt to determine the relative importance of each of the alternatives. This process is akin to partitioning the variance in ANOVAs that is due to each of the working hypotheses. For instance, Rick observed that spittlebugs, plume moth caterpillars, and thrips all fed on seaside daisy along the California coast. He wanted to

---

Box 5. *Generating alternative hypotheses*

Once you have identified a pattern that is interesting to you, think about a working hypothesis to explain or produce that pattern. Next consider alternative hypotheses that could also produce that pattern. The following list of possible factors might get you started:

· abiotic factors (precipitation, temperature, light, fire regime, etc.)
· predators, parasites, and disease
· mating factors (sexual selection, nest-site availability, opportunities for offspring, etc.)
· microhabitats (shelters from abiotic conditions, predators, etc.)
· disturbance attributed to human influences or natural causes
· genetic or ontogenetic (developmental) influences

Your list of alternatives can get long and unwieldy, but this is an important step in doing good science. You don't necessarily have to test all of your alternatives, although getting them all down on paper for consideration is a first step. Prioritize them based on how compelling and how testable each one is, and begin with your best bets.

---

know how those three herbivores affected each other. Instead of just testing the hypothesis that they competed, he examined the relative importance of interspecific competition, predators and parasites, and plant genotype on the success of each of these common herbivores (Karban 1989). This was done by including all three factors (competition, predation, host plant effects) in one experiment and partitioning the variation in herbivore performance (survival, fecundity) that could be attributed to each factor.

This experiment considered three different factors that could affect herbivore performance, but it only examined the effects of complete removal of two of those factors, competitors and predators. In other words, it compared the effects of the complete removal of competitors (or predators) with natural levels of competitors (or predators). In the jargon of statistics, there were only two levels, all or none, and each of these levels was replicated 30 times. This design works best if the effects of the predictor (in this case, presence/absence of competitors) on the response variable are linear. Unfortunately, ecological effects are often not linear. Examining the natural relationship between numbers of competitors and performance over space or time can provide valuable intuition. When you suspect that the relationship between the predictor variable that you are manipulating and the response variable may be nonlinear, you can adopt a design that involves many levels of the predictor variable (Cottingham et al 2005). For example, you could assign many different levels of competitors that span the entire range of values you observed in nature. This design would be analyzed with a regression rather than an analysis of variance (ANOVA). Regression is similar to ANOVA except that it has many levels (rather than two or a small number in ANOVAs) and it does not require replication at each level as ANOVAs do. Regression designs need not assume a linear relationship between variables, and allow you to determine the shape of the relationship. However, regression with many levels works best with one or a few predictor variables; it becomes unwieldy when multiple factors are examined simultaneously. In addition, it may be difficult to set up experimental treatments with many levels of predators or competitors. Other treatments, such as fertilizer application, lend themselves more readily to experiments with many different levels.

Recently biologists have become interested in Bayesian statistics, in part because they allow us to evaluate how well multiple working hypotheses fit data (Hilborn and Mangel 1997, Gotelli and Ellison 2004). The result of a Bayesian analysis is not a rejection of a null hypothesis, but rather an index of confidence in each of several hypotheses. A Bayesian approach lends itself beautifully to evaluating alternatives and has the potential to be a valuable tool for ecologists capable of using it. Unfortunately, Bayesian analyses require more computational sophistication than the methods that field biologists have traditionally used, and "ecological detectives" without a lot of background and confidence in mathematics may find them inaccessible at this point. No matter what statistical techniques you use, experiments that allow you to test multiple hypotheses will often be more effective in ecology that those that reject single null hypotheses.

Negative Results

Earlier we discussed the statistical procedure that lets you reject null hypotheses with varying levels of confidence. When we fail to reject a null hypothesis (i.e., when $p > 0.05$), does that mean that the null hypothesis is true? In other words, if we ask whether two populations are different and we find that we cannot conclude that they are different at the 0.05 level, should we conclude that they really are the same? The answer to both questions is no. Based on the information we have, all we can conclude is that we failed to find the difference or effect we had hypothesized. Our statistical tests give us far more power to reject hypotheses than to accept negative results as reality. In most cases we have very weak power to evaluate whether two populations are similar. Furthermore, we rarely use statistics to address this question. Many negative results (by which we mean results that are not statistically significant) in ecology never get published, a loss to the scientist who did the work and to the ecological community that never got to hear about it.

Techniques are available that allow you to evaluate whether the effect caused by one factor is as great as the effect caused by another factor. These techniques allow you to learn from negative results and make them nearly as useful as "positive" results. Unfortunately, larger sample sizes are required to reliably infer that a treatment failed to affect a population than to conclude that it did have an effect (Cohen 1988). Our ability to accept a negative result depends on the effect size (the degree to which the treatment means were different). By convention, some statisticians tend to define a small effect as a difference of 0.10 (10%) or less and a large effect size as 0.40 (40%) or greater (Cohen 1988). Imagine you are testing for an effect and find no evidence of one. You can be more confident that you didn't miss a large effect but less confident that you didn't miss a small effect, because small effects are harder to detect. Cohen (1988) provides a very readable discussion with worked examples of how to calculate the probability that your result of "no significant difference" reflects the actual biological situation at your study site. Many stats packages calculate a value for statistical power to detect effects; the value tells you how likely you were to find a significant difference given your amount of replication and your observed effect size (whether small, large, or in between).

It is often informative to compare a factor that didn't show a significant treatment effect to one that did. You can then calculate the probability that the negative effect (that is, the one you didn't find) was not as large as an effect that you did find to be statistically significant. For example, Rick tested the hypothesis that early herbivore damage, which reduced survival of later herbivores on wild cotton plants, would also increase growth of the plants (fig. 5, Karban 1993). He failed to find this effect of induced resistance on plant growth. In fact, early damage tended to reduce plant growth, although these effects weren't statistically significant. Had he missed effects that were real? He could be reasonably sure (99% confident; all confidences calculated using techniques in Cohen [1988]) that he hadn't missed a large effect (that the means were 40% different from equal). However, he was much less sure (only about 20% confident) that he hadn't missed a small effect (that the means were 10% different). These values are a little difficult to put into perspective. In the same experiment, he found that intraspecific plant competition reduced plant growth. By comparing the effect size due to plant competition and

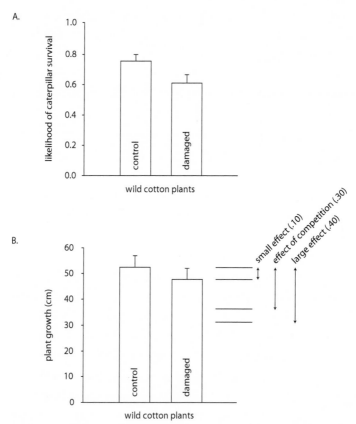

FIGURE 5. Effects of early herbivory on the success of caterpillars and the growth of wild cotton plants (Karban 1993). Early damage to the cotton plants by herbivores ("damaged") significantly reduced the likelihood that later caterpillars would survive (A). However, the study failed to find a significant effect of early herbivory on plant growth (B). In a simultaneous study, plant competition significantly reduced plant growth by 30% (histogram not shown). The lines to the right of histogram (B) show the effects of early damage that would be expected based on a hypothetical large effect size (0.40 difference in means) or small effect size (0.10 difference in means), and by an effect size equal to that empirically found to be caused by plant competition. These comparisons allow us to compare the effect

the effect size due to induced resistance on growth, he could be 30% confident that effects of induced resistance were not as great as those of plant competition. Calculations such as these allow us to report both "positive" and "negative" results that are not statistically significant and to compare the relative importance of different effects. Considerations of negative results should receive more attention from ecologists, especially those who work in systems where large sample sizes are possible.

A useful way of comparing effect sizes from different studies is meta-analysis (Gurevitch and Hedges 2001). In meta-analyses, each study becomes one independent measure of the effect of a particular factor or response variable. For example, we might be interested in the effect of removing top predators on lower trophic levels in a variety of published studies. A meta-analysis lets us formally and statistically compare the results of many studies, and lets us put our own experimental results in a much broader context. For each published study, we can calculate an estimate of the effect size by comparing the means of the treatment groups (e.g., with and without top predators), standardized by the variance. (See box 4 for more on how to calculate effect sizes.) The meta-analysis can be used to evaluate

size that we found for early damage with a defined standard (Cohen's [1988] large and small effects) and with effects of other factors (plant competition) actually found in this study. For example, this examination allows us to report, "Early damage did not produce a large effect on plant growth and was likely less important than plant competition."

whether conclusions from one study are general. It can also provide information about the conditions under which extrapolation from our experimental results is warranted. For example, meta-analysis revealed that studies of trophic cascades (removal of top predators) produced larger effects in aquatic systems than in terrestrial systems (Shurin et al. 2002). Clearly this general conclusion would not have been possible based on the results of any single experiment. Meta-analysis is meaningful only if it includes all available information on a particular question; this is another reason publishing negative results can be critically important to advancing the field.

CHAPTER 5

Using Surveys to
Explore Patterns

Earlier we discussed why manipulative experiments are such powerful tools for establishing cause-and-effect relationships (chapter 3). But in practice, even when experiments are possible, they can test a limited number of causal factors. Experiments will rarely be easy in, for example, conservation or global change biology (Young 2000). As ecologists attempt to address environmental problems with policy implications, such as issues involving introduced or endangered species, it is often impossible or unethical to conduct manipulative experiments. Luckily, observations can also allow you to pose and evaluate hypotheses.

Observations of patterns form a continuum ranging from poking around to formal surveys. Surveys are observations that are conducted in a systematic way. Surveys allow you to observe patterns at the scales at which they naturally occur, and they can provide insights about relationships that you might not otherwise notice. Less work is required to survey a factor than is required to manipulate it. So, you can realistically survey many more variables than you can manipulate, and you can observe patterns with less of an *a priori* understanding of the mechanisms producing those patterns. That said, the analysis of observational data is often less straightforward than the analysis of experimental results, and the inferences that can be drawn are generally not as strong. In the next section, we provide some suggestions for forming hypotheses and analyzing and interpreting surveys.

Forming Hypotheses from Observations

Surveys allow you to look at the relationships among many factors. Nonetheless, it is still very important to formulate explicit hypotheses about the relationships or interactions and to be clear about the questions that you are trying to answer. What do you ultimately want to know? An explicit question will enable you to identify your response variables or the outcomes of your hypothesized causal paths as well as the relevant spatial and temporal scales. We recommend that you start by writing down the mechanistic paths that you think could be possible (without worrying about what you can actually measure).

In the example involving deer and caribou shown in figure 1 (chapter 2), there were two different paths (hypotheses) that could lead to the negative effect of deer on caribou: deer could depress shared food levels (fig. 1C) or increase levels of a shared parasite (fig. 1D). Of course, a closer look at this system reveals many other factors that are potentially important for caribou, including anthropogenic factors such as road building and logging, the predominant vegetation type, other competitors such as moose, and predators such as wolves (fig. 7, pp. 93–95, from Bowman et al. 2010). Ultimately, our goal is to identify the factors that are important for caribou. Some of the paths are more likely than others, and some can be eliminated completely. For example, logging can affect the abundance of deer and moose by altering the cover of deciduous trees (an important food source), but deer and moose are very unlikely to affect the amount of logging.

Scope and Randomization

Surveys allow you to observe nature at much larger spatial and temporal scales than manipulative experiments do. Since extrapolating beyond the scope of your data can be problematic, it is a good idea to conduct your survey at the scale you care about. For example, Walt Koenig and collaborators have been interested in explaining the boom-and-bust production of acorns by California oak trees. They hypothesized that weather patterns may cause some of the year-to-year fluctuations that they observed at their initial field site in the Carmel Valley. At this field site, temperatures in spring predicted the size of the acorn crop (Koenig et al. 1996). Is this link limited to this one field site, or does it hold over larger scales? They expanded their survey and found a similar relationship between spring temperatures and acorn production throughout California (Koenig and Knops 2013). At their initial field site, trees were found to produce large acorn crops prior to years of heavy rainfall, a pattern that favors seedling establishment but is an implausible behavior. However, this pattern did not hold at other sites in California and cannot be considered a general pattern (Koenig et al. 2010). Indeed, different factors appear to limit oak seedling recruitment at each site, making extrapolations over space questionable (Tyler et al. 2006). In this example, surveys were conducted at multiple sites over the scale of the entire region, since that was the scope over which they wanted to draw conclusions. Without surveys over multiple sites, they would have made erroneous generalizations.

When selecting sampling units for surveys, you want to achieve an unbiased sample of observations across the scope of interest. This can require some thought. For example, if the scope of your study is the grasslands of the Pacific Northwest, it may make sense to get a map of grassland communities in this region and randomly choose some fraction of them to survey. However, if the hypothesis that you are testing asks whether different plant communities are found on harsh serpentine soils than on non-serpentine soils, this type of randomization will be inappropriate, since serpentine soils are uncommon, and a random sample of the region would likely include very few serpentine sites. In this case, it probably makes more sense to randomly choose a set of serpentine sites, and also randomly choose a set of non-serpentine sites. As with manipulative experiments, it is important to intersperse sampling units (sites) that have different explanatory factors (Legendre et al. 2002). In the serpentine example, you would not want all of your serpentine sites next to each other. Another reasonable strategy would be to choose a random sample of serpentine sites and then find a non-serpentine site near each of these that were matched to the paired serpentine site in other respects. However, this limits the scope of your inference about non-serpentine grasslands to those few patches near serpentine sites.

In designing an unbiased sampling protocol for a survey, it's important to invest effort in finding samples that are independent. Intersperse your samples across other factors that could otherwise make outcomes non-independent. Shared evolutionary history is one important type of non-independence between species. Perhaps you want to compare the feeding habits of lizards in desert scrub surrounding the Gulf of California with lizards in similar habitats surrounding the Mediterranean. You might observe that many North American lizards have prehensile tongues and often eat insects. Lizards from the Mediterranean lack these traits. Since many of the Californian species are related iguanian lizards and these traits (prehensile tongues and insectivory) are generally restricted to this group, a sample of the dozen or so species from each location does not provide you with independent data points from these two environments (Vitt and Pianka 2005). Any differences in the feeding habits of lizards from these two locations probably results from the shared evolutionary histories (ancestors) at each site rather than from the current ecological conditions. When comparing among species, it is possible to take shared phylogeny into account. For example, when Anurag Agrawal and Peter Kotanen (2003) designed a study to compare how much herbivory native versus nonnative plant species experienced, they chose pairs of species within the same genus or family where one was native and the other was introduced. In this design, each taxonomic group was equally represented in both treatments.

It is not always possible to completely intersperse sampling with respect to all of the factors that could cause non-independence. A number of statistical techniques remove some of the effects of non-independence among samples. Multiple regression calculates the effects of several different factors on the same response variable. Spatially or phylogenetically explicit regression factors out the effects of shared (non-independent) geography or ancestry. We give an overview of how to apply these statistical techniques later in the chapter. However, even the fanciest statistical tools cannot calculate away complete non-independence of samples, and more independent samples will provide results that are more likely to provide meaningful inferences. On the other hand, finding that a pattern has a strong signature due to a spatial or phylogenetic constraint may be interesting in its own right. In any case, it is important to think about the independence of your sampling units and to design your study accordingly.

How Many Observations?

The number of independent observations that you make in your survey will determine the size of the effect that you can detect, the number of factors that you can relate, and the likelihood that you will ever have a weekend free. So, if you care about effects that are likely to be small (weak relationships) or interactions between many variables, you will need more samples. For example, you will need fewer observations to find a relationship between soil nitrogen and tree growth than to examine relationships between soil nitrogen, phosphorus, magnesium, and calcium and tree

growth. Another way to make a hypothesis more complicated is to test for nonlinear relationships between factors.
For example, if you hypothesize a positive linear relationship between nitrogen and tree growth, you are likely to
need fewer samples than if you hypothesize a relationship
that plateaus or is hump-shaped.

As a rule of thumb, Ian shoots for a sample size of 60
when he is conducting a correlative survey. During the
planning stages he goes through a mental list of the points
listed above. How strong or weak a relationship would still
be interesting to him? If he expects, or only cares about,
a strong relationship, he might reduce the number of observations to, say, 40. How many explanatory variables does
he want to consider, and are their effects likely to be linear? If he is trying to relate three factors and one is likely
to be nonlinear, he might increase the number of observations back up to 60. Is it physically possible to make 60
independent observations? If the answer is no, he might
gamble with 30. As the three of us discussed this issue, we
found that sample sizes in various observational studies
that produced interesting results ranged from nine (a very
clear and simple relationship where each sample was expensive and required a year-long survey) to many hundreds
(a weaker, more complicated relationship).

What Factors to Measure?

The number of factors that you record for each replicate is also important to determine. The simplest scenario
is to record and relate two factors, where only one of the
factors can affect the other. For example, you could relate

the breeding time of a bird species with the average March temperature at multiple sites (replicates). It is reasonable to hypothesize that temperature affects bird breeding but not that bird breeding affects temperature (fig. 6A). This model could be made more complicated by adding more predictive factors. In addition to temperature, March precipitation might be important, and temperature and precipitation in other months could play a role as well. As the number of explanatory factors increases, you will need a greater number of observations to examine relationships between these factors and a response variable (in this case, bird breeding). At one extreme, it is impossible to relate more factors than you have independent observations; if you wish to explore 20 possible explanatory factors, you will need more than 21 independent observations (many more). Another complication comes about when you are interested in the effect of a factor on more than one response. For example, you might hypothesize that March temperature affects both the timing of bird breeding and the activity of snakes that eat bird eggs. Don't go around recording things willy-nilly; make sure that the factors you measure relate to your interests.

There is some disagreement about whether it is better to concentrate your survey on a single well-honed hypothesis or to design the survey to consider relationships among many factors that might not be obviously related. When we start a survey, we attempt to have a clear, simple hypothesis in mind, but we also try to be flexible and opportunistic. If there are other factors that might be important and easy to measure, why not keep track of them? Of course, there are limits to what you can measure, and you don't want to dilute your efforts by attempting to keep track of factors that you will not use in your final model. For example, Rick has surveyed a population of wooly bear caterpillars at his field site for the past 30 years (data are available at http://karban .wordpress.com/ltreb). Since they are heavily attacked by a tachinid parasitoid, he also recorded rates of parasitism by these flies, but parasitism has not proven to be a good predictor of caterpillar success or numbers (Karban and de Valpine 2010). Recent results indicate that ants can be important predators of caterpillars and that ants and rodents can greatly reduce numbers of pupae (Karban et al. 2013; P. Grof-Tisza, unpublished results). Rick sorely wishes he had recorded some quick and dirty estimate of ant and rodent numbers during each of these years. However, there are so many factors that could potentially affect numbers of wooly bears (abundance of host plants, diseases, birds, other predators, measures of potential host quality, and so on) that it would have been prohibitive to keep track of all of these factors.

Analyzing Survey Data

Conducting a survey takes less time, effort, and cost than setting up experimental treatments and allows you the immediate gratification of putting meaningful numbers in your notebook. Much of the tricky work comes in the analysis of survey data, though statistical techniques are becoming increasingly available that aid in (1) disentangling the relationships among multiple factors measured in a survey, (2) accounting for confounding patterns in the data such

A. March temperature ⟶ bird breeding

B. March temperature ⟶ insect emergence ⟶ bird breeding

C. March temperature   bird breeding
              ↖   ↗
                  day length

FIGURE 6. Possible relationships between March temperature and bird breeding. A. Variation in March temperature causes variation in bird breeding. B. March temperature causes a change in another intermediate factor, insect emergence, which then causes a change in bird breeding. C. A third factor, day length, causes changes in both March temperature and bird breeding, although temperature and bird breeding are not causally related.

as spatial and phylogenetic correlations, and (3) gaining insight into cause-and-effect relationships among variables.

One frustration with conducting surveys is inferring causality with confidence. For example, finding a positive correlation between temperature and bird breeding does not necessarily mean that these two factors are directly linked (fig. 6A). Temperature could potentially affect the timing of bird breeding through an intermediate factor, perhaps the emergence of insects that the birds eat (causal but indirect; fig. 6B). Alternatively, a third factor (length of daylight) could potentially affect both temperature and bird breeding (noncausal; fig. 6C). If it is possible to experimentally manipulate the factors (temperature, insect emergence, day length) independently, we can be more confident about which ones cause which others. These manipulations are not always possible. We can use correlations to assess whether two factors are related when there is no hypothesized causal relationship between the factors. When we have an *a priori* idea about the direction of cause and effect (e.g., weather drives bird breeding), we can use regression analysis to relate the effect of the predictor on the response. The response (*y* axis) is described as a mathematical function of the predictor (*x* axis), and the relationship between the factors is displayed as the trend line on a scatterplot.

DISENTANGLING MULTIPLE FACTORS

In many cases there are multiple factors that could cause a response, and it can be difficult to disentangle these potential drivers. For instance, Ian could ask which leaf traits affected the success of caterpillars that fed on the leaves. In his system, it is difficult to manipulate many of the leaf traits in any meaningful way—how do you make leaves tougher or hairier? However, there is considerable natural variation in leaf traits among the oak species he studies. So Ian could conduct an observational study in which he measured multiple leaf traits on numerous oak species. He could then place caterpillars individually on leaves and measure their weight at pupation. With these data, he could assess the relative importance of multiple predictive factors (leaf traits) on a single response (pupal weight).

In the simplest scenario, he could use regression analysis to separately assess the relationship between each leaf trait and weight at pupation. He could plot (or regress) each explanatory factor against weight at pupation and also examine the correlations between the leaf traits. However, because he did not independently manipulate each of the predictor variables, many of them are likely to covary (be correlated). For example, Ian has found that leaves that are tougher also tend to have higher concentrations of phenolic chemicals. Part of the statistical effect of toughness actually could be caused by phenolics, or vice versa.

Multiple regression is a technique commonly used to divvy up the relative importance of shared (collinear) effects of multiple predictor variables on a response. Multiple regression estimates how well the whole model (i.e., the combination of all predictors) describes the response, and how well each predictor relates to the response after accounting for their collinearity with other predictors. When two or more predictors are highly correlated, it is difficult to distinguish the effect of one predictor from another. In this case, the two predictors will vie for the shared effect, producing erroneous results. You can get some sense for whether this is a problem by plotting the correlations between predictor variables. If they are not obviously related to one another, there is likely no problem. (The potential correlations can be quantified by calculating the variance inflation factor [VIF] for each predictor variable [Sokal and Rohlf 2012]. If two predictors are highly correlated, VIF values will be higher; VIF values greater than 10 indicate that predictors are too highly correlated to distinguish their separate effects.) If some of the predictors are conceptually related, perhaps one can be removed. Another strategy is to look only at the shared portion of the predictor variables to determine something about their shared effects on the response. Principal component analysis (PCA) provides a view of the covariation between two or more variables. In essence, PCA condenses multiple factors into those portions that have the most covariation. By using the condensed PCA axis as a predictor, you assess the joint effects of multiple variables on the response. The downside of PCA is that the new condensed variables are often not easy to interpret.

Models that include many different potential predictors will explain more of the pattern but may be more complicated than informative and end up both hard to interpret and unlikely to represent a general pattern. As ecologists gain access to larger data sets, it becomes increasingly possible to examine the relationships among many factors. In these cases, the challenge of the analysis becomes separating those few meaningful factors from many others that might not be important. When there are many predictor variables, some might mask the effects of others even if there is not excessive collinearity between any pair. Several information criteria measures (including Akaike Information Criteria [AIC] and Bayesian Information Criteria [BIC]) have been developed to help remove predictors that are not adding accuracy to predictions (Burnham and Anderson 2002). These statistics allow you to choose a set of relatively few predictors, making for a simple model that creates an accurate description of the response. Dangers of this approach are that you may throw out a factor that is actually meaningful and that you may erroneously include a factor that appears significant by chance alone (recall the relationship between storks and babies, fig. 4).

NON-INDEPENDENCE AMONG OBSERVATIONS

In designing experiments, you can randomly assign treatments so that the replicates are interspersed (chapter 3). But in observational studies, it is sometimes not possible to have replicates with interspersed factors. Statistical techniques have been developed to deal with two forms of non-independence of samples—spatial clumping and phylogenetic clumping.

Two observations that are close together in space are less likely to be independent because the same single event may have affected both of them. For example, if two sea anemones within a few centimeters of each other are sampled, it is possible that the same butterfly fish grazed on both of them. Spatial autocorrelation exists when samples that are close to each other are more likely to share circumstances than those farther apart. If you know where a sample is located, it is possible to test for similarity based on distance and to remove this effect using a spatially explicit generalized least squares (sGLS) analysis (Legendre et al. 2002, Dray et al. 2012). This is a form of multiple regression that tests whether samples that are close together might also be similar to one another for reasons that have nothing to do with the predictor variables. For example, perhaps you wish to determine the survival rates of individual anemones at multiple sites. If similar outcomes occur to samples that are spatially close together, they might not be independent. Using sGLS allows you to deemphasize events that are close together in space.

Another form of non-independence between samples occurs when the samples share common ancestors. Numerous traits are shared among close relatives, whether across the tree of life or among your family members. Whether you are comparing species or individuals within populations, it is often useful to account for lack of independence due to shared ancestry (phylogeny). For example, if you wanted to compare the relationship between garlic consumption and health benefits, you could survey individuals and record how much garlic they eat and their incidence of cancer or attractiveness to vampires. However, Rick and several members of his immediate family hate garlic. Garlic aversion and health probably both result from shared genes and are probably not causally linked to each other. Individuals (or larger groups) that share common ancestors are unlikely to be independent samples.

There are many situations in which including information about shared phylogeny can be informative for observational studies. Phylogenetic comparisons allow you to identify broad-scale patterns across many taxa over long periods of time (Weber and Agrawal 2012). Manipulative experiments will rarely be able to provide comparative information over broad taxonomic and temporal scales.

Phylogenetically explicit generalized least squares (pGLS) analyses and several related techniques take into account the non-independence of shared ancestry (Garland et al. 2005). These analyses require an accurate phylogeny and remove the structure of evolutionary relationships from the model. Ian used this method in his analysis of relationships between leaf traits of 27 oak species and caterpillar performance described above (Pearse 2011). Relationships among the oak species had been determined in a previous study (Pearse and Hipp 2009), and pGLS was used to factor out the effects of shared ancestry. Fortunately most of the plant traits (e.g., tough leaves) that caterpillars cared about had evolved multiple times. If there had been few independent origins of those traits, the study would have had little ability to differentiate effects due to shared history from effects due to leaf traits.

PATH ANALYSIS AND CAUSAL RELATIONSHIPS

Path analysis is a technique that allows a researcher to consider different causal schemes (paths) and to evaluate which of those paths does the best job of explaining the observed patterns, that is, which factors are likely to cause which other factors (Shipley 2000, Mitchell 2001, Grace 2006). The diagrams with arrows that connected weather, plant stress, and herbivore numbers in chapter 2 are path diagrams. The easiest way to evaluate paths is by using partial regression coefficients to estimate the strength of individual paths (arrows in the diagrams) along with goodness-of-fit statistics to estimate the fit of the overall model (see Mitchell 2001 for a very clear description with worked examples, and Grace 2006 for a more complete, but still accessible, account). Partial regression coefficients provide an estimate of the effect of one variable on another when the effects of other variables are held constant.

Path analysis can shed light on the likely validity of several different cause-and-effect hypotheses (Shipley 2000, Mitchell 2001, Grace 2006). It is particularly useful for evaluating the roles of indirect effects (fig. 7).

Path analysis that includes structural equation modeling and maximum likelihood methods can be used to generate plausible hypotheses when several are possible, to estimate effect sizes, and also to confirm that one causal hypothesis is more likely than others. It can be particularly valuable when there are many factors that could potentially interact in a variety of different ways; in such cases it may be difficult or impossible to experimentally manipulate all of them.

Path analysis can be extremely useful early in a study before you commit to time-consuming manipulations, and

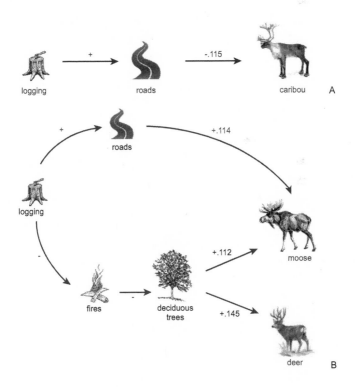

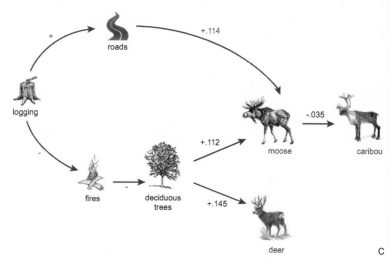

FIGURE 7. A path analysis based on results from Bowman et al. (2010) showing various hypotheses about effects of logging, roads, fire suppression, occurrence of deciduous trees, and abundance of wolves, deer, and moose on numbers of caribou in northwestern Ontario (see fig. 1 for some natural history of this system). The authors surveyed 575 cells, each 100 km². A. Human activities, including logging and associated road building, had negative impacts on the number of caribou. These effects were found to have a significant but weak overall effect on caribou numbers. B. Logging and fire suppression had positive effects on the prevalence of deciduous trees, and deciduous trees were associated with increased populations of deer and moose. C. Roads were also positively associated with moose, and a greater number of moose was associated with fewer caribou. As a whole, these anthropogenic activities increased numbers of deer and moose. A greater number of moose in particular was associated with fewer caribou (effect of deer not significant and not shown). D. These anthropogenic changes also increased the density of wolves, and wolves further decreased the number of caribou. The path coefficients give an estimate of the strength of each of the effects and allow us to calculate the relative direct and indirect effects of various

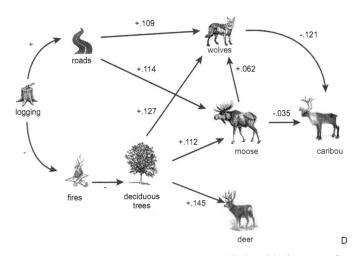

ecological mechanisms. For instance, the relationship between deciduous trees and caribou is the sum of effects of deciduous trees
increasing moose, which decrease caribou $[(+.112)(-.035)] = -.004$,
plus the effects of deciduous trees increasing numbers of wolves,
which in turn decrease caribou $[(+.127)(-.121)] = -.015$, plus the
effects of deciduous trees on moose, which increase wolf populations, which decrease caribou $[(+.112)(+.062)(-.121)] = -.001$. Adding these three negative indirect effects together gives the sum of
the effects of deciduous trees on caribou, $(-.004) + (-.015) + (-.001)$
$= -.020$. It is apparent that the indirect path involving wolves has the
largest negative effect on caribou numbers. This analysis suggests
that the overall effects of roads and the indirect effects mediated
through changes in the wolf population are more important than
effects mediated through potential competitors of caribou, moose,
and deer. This hypothesis could possibly be tested by experiments
that manipulate densities of roads and wolves. Of course, this analysis omits other paths that may also be very important for caribou,
some of which were included in the survey but found not to be significant and some which were not considered (e.g., the meningeal
worms mentioned in figure 1).

it is particularly valuable when used along with experimental manipulations. The simple act of drawing out hypothesized relationships as paths is a good opportunity to formally think about the way you view your system. Path analysis using structural equation modeling allows you to increase your confidence when inferring causality from the results of observational studies. In cases where both experimental manipulations and observations have been conducted, they can be combined in a single path analysis.

Results from path analysis are sensitive to several assumptions. Perhaps the biggest problem with these new techniques is that they depend upon a number of assumptions about causal structure. For example, it is assumed that there is no feedback between factors but rather that causality is asymmetrically directed (A causes B, and B does not feed back to affect A). But feedback of this sort is common in ecological systems and makes inference of causality much more difficult and murky. Despite these limitations, path analysis is still a useful tool for making your thoughts explicit and for testing causal models in some cases.

In summary, observational studies are useful tools to evaluate many factors simultaneously and to generate causal hypotheses. However, conclusions drawn from observational studies are generally not as firm as those inferred from well-designed manipulative experiments. Observational and experimental approaches can address different, often related, questions.

CHAPTER 6

Building Your Indoor Skills

Unfortunately, field ecology doesn't take place exclusively in the field. In this section, we offer suggestions about skills that might make your life a little easier.

Organizing a Season

Organizing a field season requires thinking of both the big picture and the day-to-day activities. Remember that you will want to share your data as some kind of story. So ask yourself throughout whether your questions (the big picture) and your day-to-day activities are allowing you to fully develop the story your system is offering. Plan out your season in your electronic or field notebook, not just in your head. You might be tempted to begin by focusing on your methods, such as treatments and sample sizes, but resist this urge. Start instead by writing down the questions you want to answer that season. This might be a good time to try writing for 10 minutes a day or using the iterative writing technique discussed in box 3, chapter 1. Make sure you have a firm handle on these questions and that they are answerable. It is amazing how often the manipulations we conduct don't effectively and directly answer the question that we want to address. This "disconnect" can often be avoided by being very explicit about the question and then asking whether the manipulation really is the most appropriate match for that question. Consider whether alternative approaches could be informative, effective, and more efficient. Run your ideas past your major professor, lab mates, or anyone else who will give you meaningful feedback on your plans. Then do some more writing before you forget their comments.

Before you begin your experiments, troubleshoot them as much as you can. What will happen if your organisms are harder to find than you expect? What will you do if you get bad weather? Come up with contingency plans. Try to anticipate all the things that could go wrong and how you will respond to them. Again, run these potential problems and solutions by someone who knows the system (your major professor or a colleague) to make sure that your ideas are on track.

Once you start your project, let your observations direct your next steps. Go where your system wants to take you. Constantly reassess what you've done so far, how you explain the results, and what alternative explanations might be, and let this information guide your decisions for how to continue. Take stock of your progress and your preliminary results at least once a week. Writing can help with this process.

A lot of the best scientists are both very lucky and perceptive enough to recognize when they have stumbled upon something unexpected and novel. Keep your eyes open and be willing to take off in a slightly different direction. Go in the directions that you determine will be the most profitable and don't feel constrained to follow the plan that you set for yourself before you started. As you redesign your experiments and also when you write them up, develop your best story (your most interesting results) and not necessarily the question that initially got you into the system. Remember that while you may be interested in the historical development of your thoughts, your audience is more likely to want to hear your most exciting results placed in the most interesting framework, whether or not you came upon that framework deliberately or by dumb luck.

Throughout your field season, take some time to come to grips with your results and what they mean. We recommend that you do this in three ways. First, analyze your data as soon as you can. It's fine if this is only a summary of your treatment means and variances; it will still give you a qualitative picture of your findings so far. Even without access to a computer, you can and should get a feel for your data, whether your sample size will be sufficient, whether you are missing a key observation or experiment, and so on. Back at home, a complete statistical analysis lets you determine which effects are real, which experiments you want to repeat, and which new directions you want to think about. Second, use your notes to write up your methods immediately; if you don't, you run the risk of forgetting key details later. Third, we recommend that you write up your results as a preliminary paper soon after returning home. This has multiple benefits. The reading that is involved with assembling the introduction and discussion can be particularly valuable. This process lets you put your questions and your results in a broader perspective and gets you familiar with the pertinent literature. (Note to perfectionists: Don't get hung up here. It is more important to get a preliminary draft of your results than to do a complete literature review.) You often get new ideas for questions that you want to ask next by thinking about the bigger ecological issues and by seeing what other people have found in related studies. This process also forces you to reevaluate and reprioritize your questions and plans for future research in light of what you have just found. In addition, writing up your results each season gives you first drafts of your publications, and material you can use to get feedback on your work. As a rule of thumb, until you make significant progress on your first research project (that is, have a rough first draft of preliminary results), you probably shouldn't take on additional activities, because you may be using them to avoid your demons.

Writing Habits

Writing is difficult for many people, but we have learned a couple of tricks that can make it easier. Because research involves producing a final written product, you may assume it means you should do the writing at the end. We don't recommend waiting that long. Instead, write early and often. For example, you could write every day for a brief period, say half an hour, or even fifteen minutes. The amount of time isn't important; the daily routine is. Start your writing routine years before you are ready to write up your thesis. Write to generate ideas (see box 3), to process papers you have read, to document your methods, and to keep track of your observations. Charles Darwin wrote every morning and was done by lunchtime. This habit allowed him to produce more than 20 books and monographs.

Another trick we have found very useful is called the Pomodoro Technique™. (Google it if you're interested.) Francesco Cirillo, who developed this method, advocated working for 25-minute intervals, punctuated by five-minute breaks. At the time, many kitchen timers in Italy, where Cirillo lived, were shaped like tomatoes (*pomodori*), which inspired the quirky name. You can experiment with the time—a popular variant is 50 minutes of writing with a 10-minute break. We have occasionally been invited to do this technique with groups of UC Davis grad students and have found it fun and productive. We hope that doing small bits of writing as you go will help demystify the process.

Learning to Read, Again

Writing is only one side of communicating as a professional; reading is the other, and the two are interdependent. By the time you get here, can we really have anything to say to you about reading that you don't already know? Reading the academic literature is not the same as reading novels, blogs, or textbooks, and what works in college won't necessarily work in grad school and beyond. As a new grad student, Mikaela was trying to read every word of every assigned article and to highlight them in different colors, all the while zoning out from boredom, lapses of focus, and inadequate knowledge of the vocabulary. If you are new to the field, the vocabulary will come after a little time, particularly if you use a dictionary (hardcopy or electronic), so go easy on yourself about this. Knowing the structure of journal articles will also make it easier for you to focus on specific parts of a paper, for example depending on whether you want to find out where the field is currently, or what specific questions were addressed by the authors (for a sense of the structure of journal articles, see box 6, "Journal article checklist," and related text in chapter 8).

One problem that all ecologists (new and old) face is the overwhelming quantity of information available. It is not possible to read all of it, but becoming overwhelmed and reading little is not a good strategy. The key is deciding on the topics that are interesting to you and then identifying the important papers on those topics. Review articles often provide an excellent entree into a field. Look for papers with "review" in their titles or those published in one of the *Annual Review* or *Trends* journals. The introductions to research papers generally cite the important papers in a field and can provide a starting place. You can also enter keywords into the Web of Science database to gain a foothold, or use this database to see which other articles have cited an article that you find particularly relevant. Don't get distracted; most of these won't be helpful. Your advisor or other mentors may also help you get started.

You may also want to consider how you read. First-year law students have been the subjects of research about reading strategies that we think also applies to ecologists. The most common reading strategies have been categorized as "default" (summarizing, paraphrasing, underlining text), "problematizing" (considering the author's purpose, posing hypotheses and questions as you read), "rhetorical" (connecting the reading to your own personal interests), and "distracted" (an inability to concentrate or focus) (Deegan 1995). Students in the bottom quartile of their class were most likely to rely on default strategies. We suspect this strategy won't work for grad students either. Students in the top quartile after their first year were most likely to rely on problematizing strategies, probably because problematizing works well in classes. The groups entered law school with no significant differences in their undergraduate GPAs or LSAT scores, so success as an undergrad did not guarantee that students possessed the reading skills required to do well in law school.

Of the four reading strategies, the rhetorical strategy is probably the most valuable for building a successful research project. Before reading a paper, decide what you want to get from it. Will it provide new ideas about research directions? Do you want to know what other people have been working on in this area? Are you interested in a technique that the authors used? Give yourself a time limit to get the information that you are looking for—get in, get what you want, get out. Allow yourself to enjoy reading within these boundaries. How do the authors justify their work? What are their main results? How do they relate their work to the bigger picture? You can think about these questions for a few minutes, but keep in mind what you came for. Write a few sentences in response to the question that you wanted to answer before you started reading the paper. This keeps you from using reading to procrastinate and forces you to deal with the material and to write. This strategy will help you make sense of your reading now and will make the writing process easier as you tackle it.

Students are often puzzled by the rhetorical strategy because it is not very critical. There is a role for critique (the problematizing strategy), since it can suggest what is missing or could be improved. But many grad students get stuck looking for the flaws in the papers they read. Every paper has them, but finding flaws doesn't necessarily justify discarding the work and may distract from the useful contribution that the paper can provide. Resist the temptation to be a critic rather than a producer.

These same strategies may be helpful when you are listening to seminars. Thinking about how the research relates to your work or preparing yourself to ask a question (whether or not you actually ask it) can make you a more attentive and less passive listener.

You have been reading for quite a while now, and it may take some time to train yourself to read in a new, more efficient way. We hope this encourages you to analyze your own reading strategy and talk to your colleagues about strategies to get through the mountain of literature we are all facing.

As you plan your activities, keep in mind that your goal is to conduct research that will teach you about nature and to communicate your results. When allocating your time, think long and hard about those activities that won't result in publications—do them only if they will make you happier or make you more employable over the long term.

CHAPTER 7

Working with People and Getting a Job in Ecology

When we aspired to become biologists, we imagined that the scientific process was totally objective; the truth could be separated from lesser hypotheses in a manner that was removed from social interactions. The longer we stay in this business, the more we are struck by the opposite. Science is a social endeavor, most ecologists are human beings, and coming up with good ideas isn't enough; successful ecologists need to be persuasive about the value of their ideas.

In this chapter we consider some of the social situations that you are likely to encounter as a graduate student and as a professional ecologist. Graduate training takes place in universities, institutions that were established during the Middle Ages and that cling unconsciously to medieval European conventions. These include what Damrosch (1995:18) described as "the indentured servitude of graduate student apprentices and postdoctoral journeymen." Perhaps because universities originated from monastic traditions, a high degree of zealous dedication and self-discipline is expected. Any less is likely to be met with disapproval. Many administrators at universities talk about "work-life balance," but the academics that you encounter (your major professor, search committee members, etc.) may have no such balance in mind.

How to Pick a Major Professor

The prestige of the university granting your degree is far less important than the level of intellectual, emotional, and financial support that you receive along the way. You should try to get a support package that will keep you from needing outside employment, but you probably shouldn't pick a program based on a few thousand dollars more or less. Grad school is a mistake if your goal is to get rich; you can increase your income by doing just about anything else. But grad school may be a great place to be if you are passionate about ecology. And above all, when you are choosing a graduate program, pick a professor rather than a university.

If you are not yet a grad student, you will probably underestimate the importance of the relationship with your major professor. However, it will color everything that happens to you, so choose carefully. You should contact potential major professors when applying to programs. Resist the temptation to explain that you've always been interested in nature and that playing with bugs has been your passion since the age of three. Instead, be prepared to explain why working with that major professor will be a good match for your interests. This person will be more impressed if you discuss actual research he or she is doing than if you say something generic about wanting to study vertebrates or work in marine systems. And later, when it comes time to submit a statement of purpose with your application, this potential major professor and the other committee members will expect it to focus solely on your research interests.

Make sure that you can communicate well with your major professor. The reputation that your major professor has as a researcher is important, but not nearly as important as his or her reputation as a mentor of graduate students. Visit potential professors before accepting any offer. Ask how long your leash will be. Are students expected to put in face time from eight to five, or do you set the schedule that allows you to be most productive? Will you be given a project, or will you come up with your own? Ask about what has happened to previous students. How many have finished their degrees? If they didn't finish, why didn't they? What jobs did they get? Correspond with past students. Would you like to be in their shoes five or ten years down the road? Professors have track records as mentors, and you can expect to face many of the same situations and eventual successes as past students.

One subject that grad students and major professors often butt heads about is authorship. Each subdiscipline has a slightly different tradition about what constitutes authorship, with lab-intensive programs including the major professor as an author more often than field-oriented programs. It's a good idea to ask former and current students about joint publications with your potential future major professor. The Ecological Society of America has developed a policy statement for authorship. To be considered an author, the person must make substantial contributions. Simply providing funding does not confer authorship privileges. Even if it is awkward, discuss authorship (including name order) before designing a study. You may want to keep this fluid, but having a discussion early on can reduce the chances of big disappointments later. These authorship guidelines apply whether or not your potential coauthor is your major professor.

Interacting with Other Ecologists and Beyond

Despite the myth, graduate school is not a completely independent pursuit. Your work as a grad student includes exams and a thesis that involve committees of professors. You will be collaborating on research, attending conferences, and enlisting the help of a wide range of people in setting up lab and field work. Success in the field of ecology will ultimately depend on these interactions with others.

Completing grad school often requires assembling at least two committees—one for your qualifying exam (orals) and the other to assess your thesis or dissertation. Involve your committee members so that they can provide advice and support much as your major professor does. Choose committee members who are going to give you the most help. Get to know them by taking their seminars, attending their lab meetings, talking to them about your project. Try to use the same people for your exams, thesis committee, etcetera, so that you know each other well. The better they know you and the more invested they feel in you, the more help they are likely to give you when it comes time to submit your manuscripts for publication and apply for jobs. Don't hesitate to run your research proposals, grants, and manuscripts past your committee members. If they are too busy, they'll tell you, but at least they will know that you are trying hard. If they are frequently too busy, you should also think about finding committee members who will invest more time in you. Don't make the mistake of "getting lost" interacting only with your own lab—you will need relationships with other professors to succeed.

If you intend to become an ecologist at a research-oriented university, your committee members are also likely to be a good source of advice about career ideas and job materials. If you plan to seek a job with a focus on teaching or policy rather than research, these same committee members may or may not "get" your vision, depending on their own biases about jobs outside of research universities. Also, even those who have the best of intentions usually will not know as much about the culture of your preferred career path as people who are already on that path. Some grad programs will allow you to add a committee member from outside your university, and this may be a good choice if you have a clear idea of the job you will seek when you finish.

As a grad student or as a professional ecologist, you can often expand the scope of your research by collaborating with other scientists. Good collaborations, like other mutualistic interactions, generally involve organisms that can offer skills or expertise that their partners cannot easily acquire on their own. Perhaps one person is a good chemist while another has little expertise in chemistry but knows a lot of natural history. Together this team is capable of going places that neither person could go alone. Sometimes people who have similar interests but different personalities are able to work together very effectively; an idea person who has trouble finishing projects can work well with a pragmatist who is a little less creative. The main disadvantage of collaborating is that you lose some control over the content and pace of the work. For example, you may be in more (or less) of a hurry than your partner, which can be frustrating to both of you. Despite this constraint, the advantages of collaborating are great. Thirty years ago, most papers in our field were single authored; today the rugged individualist who works completely independently is relatively uncommon.

Another valuable way to interact with colleagues is by attending professional meetings and conferences. Even if you find these events stressful, it is worth making yourself go and trying to interact as best you can. Be kind to yourself at meetings by not having unreasonable expectations and by pampering yourself. (For example, take a nap or go for a walk if you are feeling fried.)

Meetings give you the opportunity to find out what other people are up to, let them know what you are up to, get feedback, and most importantly, schmooze. Don't feel like you need to introduce yourself to the big dogs in your field. Any and all interactions can be beneficial. Since personal contacts are so important, getting familiar with the people in your field is extremely worthwhile and will help you start collaborations, get your manuscripts and grants accepted, and let you feel part of a community.

Back at home, research projects and careers in ecology put you in contact with a wide range of people, including administrative assistants, reserve stewards, and resource managers, among others. Unfortunately, some ecologists see the world as a caste system with academics on top. But publishing a paper requires the help of many different people. The good relationships you cultivate with the people who facilitate your work can be mutually rewarding. Even if you are shy, make an effort to communicate your respect and appreciation for these facilitators.

One way to get research accomplished is by hiring helpers. They can do repetitive work, freeing you to do more creative tasks. Research assistants can also get things done at your field site when you can't be there. In addition, some projects require a lot of hands; assistants can allow you to answer questions that you could not address as one person. However, there can be severe downsides to hiring helpers. They are expensive and have much less invested in the quality of the work than you do. For many assistants, it's just a job. Hiring helpers generally involves work like writing grants and progress reports, and this paperwork can make you the administrator and your assistants the biologists. Assistants generally don't have the expertise and intuition for the system that you do. Often when we have asked other people to do routine tasks, we have been surprised when they weren't done "right." There are many little things, "tricks" if you will, that each of us takes for granted. It is very difficult to convey all of these to another person, no matter how detailed the instructions. All of this subtle intuition comes from working with your organisms. If you hire someone else to do the hands-on work while you do the paperwork, you will miss out on developing intuition about the details of your system (and maybe about the big questions as well). One way to minimize these risks is to work alongside your helpers. That way you can provide more quality control and also develop critical intuition.

How to Get a Job in Ecology

Getting a job is one of the most important things you'll do, and something you'll want to start thinking about years before you finish your degree. Think carefully about the kind of job you would like, how to find it, and how to make your application attractive.

The first step is to do the work of figuring out what type of job you would like most. Different jobs will require additional skills and experiences beyond publishing research. It is well worth your while to ask yourself what job you would most enjoy. This question may be difficult to answer because how should you know what jobs might be available and what it may be like to do them? You may have been trained (brainwashed) to value some jobs more than others because of the prestige, money, or security they will bring. We highly recommend Beck (2001) for an in-depth and surprising treatment of this subject. She distinguishes the "social self" from the "essential self." Your social self values the things that are valued by the people around you. Your essential self knows how you truly want to spend your time because it has "compasses pointed towards your North Star"

(Beck 2001:3), but it may be repressed by your social self. Many of us subconsciously allow other people to direct our choices away from the paths we would otherwise most enjoy. It may be more difficult to untangle the influences that direct your path than you might think. For example, Mikaela entered graduate school with the expressed intent of eventually working at a small teaching-focused college. Within a few months, she was telling people she wanted to become a research professor, and she believed it herself. It took several misdirected years before she even questioned these aspirations, which were guided by her social self, not her more genuine essential self.

Who are these people who are squelching our internal desires? They are our peers, parents, and professors (not to mention ourselves)—people who, in most cases, have the best intentions for us. Disappointing these people can be very painful, but allowing them to direct your path has grave, and often hidden, consequences as well. Beck (2001) has many surprisingly enjoyable exercises to help you figure out what your essential self wants and how to get there.

Once your goals are clear to you, it is important to be strategic to accomplish them. As we mentioned at the start of this book, the currency for essentially all jobs in ecology is publications. (Jobs at community colleges may be the one exception.) Search committees for jobs that will require research as well as for those that will not all want to know what you have done in the past, and they evaluate you by looking at your publication record. Skills, insights, support, and collaborations that will help you build that record are worth pursuing. There are many sources of information about the nuts and bolts of putting together a successful job application and preparing for your interviews. We will not cover this process here but recommend Chandler et al. (2007) for a focused and relevant guide to creating a strong research CV and statement of research interests. We've found that the best information about writing teaching CVs and statements of teaching philosophy can be found at the websites of university teaching centers such as Washington University's http://teachingcenter.wustl.edu/writing-teaching-philosophy-statement and University of Michigan's http://www.crlt.umich.edu/tstrategies/tstpts.

As you become clear about what you want to do, make sure you find out about the particular subculture of the "club" you hope to join. To choose one example, if your focus is community colleges, we've been told that it can be easier to find a tenure-track position if you get experience teaching at the community-college level while you are still completing your degree. Waiting to get that experience until after finishing your PhD puts you at higher risk of becoming a permanent "freeway flyer" who gets hired for single courses at multiple schools but isn't seen as tenure-track material. Each subculture has its own unpublished rules, so it's worth asking around.

When it comes time to apply for jobs, you'll need to figure out the best places to look. Openings are often listed on the websites of *Science* and *Nature*, in the *Chronicle of Higher Education*, at the Ecological Society of America's online job board, and at Ecolog. An outstanding resource for academic jobs in ecology is http://biology.duke.edu/jackson/ecophys/faculty.htm, which is maintained by Rob Jackson.

It's worth asking people who have jobs in your chosen field if there are websites specific to your interests.

Many of the most satisfying jobs are never listed—you create them. Figure out what skills or expertise you have that would be valuable to an employer. Next, figure out who might find those skills or experiences appealing and might be in a position to offer you a job. Approach that person with your proposal and make the case that you can provide something that they will value. The popular job-hunting book *What Color Is Your Parachute?* (Bolles 2013; first published 1970) has advocated this method of creating your own job for decades. Don't let the sketches and exercises put you off—this method can be remarkably successful. Rick got his first job by approaching the provost of his undergraduate alma mater, Haverford College, with the argument that ecology should be offered and that he was the guy to do it. It seemed like a huge and scary long shot, but it worked, and we have witnessed similar successes repeated in various employment situations.

There are more jobs at teaching-focused institutions than at research universities. Most graduate students receive little training that prepares them for teaching careers (Gold and Dore 2001). If you aspire to a job that emphasizes teaching, in addition to demonstrating that you understand ecological research, you should gain teaching experience (Fleet et al. 2006). If you are getting your degree at a research university, you may well be getting the wrong idea of what will be expected of you when you go on the market for a teaching-focused job. For example, between 57% and 67% of biology faculty members at primarily undergraduate institutions expected successful applicants for faculty positions to have taught their own course, while only 34% at primarily research institutions had this expectation (Fleet et al. 2006). All of the cover letters and statements of teaching philosophy from applicants for teaching jobs say how enjoyable, valuable, and rewarding teaching is. To make these statements seem credible, you should acquire some actual teaching experience. Being a TA (teaching assistant) in a class taught by someone else won't set you apart. Instead, go out of your way to develop and teach your own course. You can make opportunities by teaching summer session courses at your institution or courses at local junior colleges. Perhaps a professor at your old undergraduate institution is thinking of taking a sabbatical leave. Several specific questions pop up repeatedly related to your statement of teaching philosophy, so consider how you will want to answer them. What do you want your students to know or be able to do? How do you help them get there? How do you assess whether you've succeeded? Finally, how do you teach all of your students, not just the ones who will succeed no matter what you do? If you are looking for new ideas to answer these questions, take a look at the research on how to teach ecology effectively. A few good sources (in order of resource abundance) are the Teaching Issues and Experiments in Ecology webpage, the CBE–Life Sciences Education journal, and Bioscience.

If you want to work for a nonprofit, for a government agency, or in the private sector, you will make yourself much more attractive if you gain some experience in that arena, and even in that organization, before it's time to apply. Many students who successfully transition from graduate school to nonacademic careers interact with scientists who already have jobs in their sector of interest (government agencies, NGOs, etc.) Making connections with potential employers while in school is valuable, and internships are another excellent way to do this. Two good sources of listings of nonacademic jobs are Ecolog-L and USAjobs.

In addition to research and disciplinary expertise, search committees for nonacademic jobs may place emphasis on interpersonal, leadership, networking, and management skills (Blickley et al. 2013). This is particularly true for nonprofits, while private-sector jobs also value technical and technological skills. Occasionally, some of the non-research-related work you do in grad school can build skills that are transferable to the job market. For example, grading papers may allow you to acquire skills that will be useful later when providing feedback to people you supervise. If you get stuck teaching a cookbook lab, try to turn the experience into an opportunity to practice presenting in front of an audience, explaining concepts clearly, making decisions with colleagues (other TAs), and so on. If possible, describe how these skills relate to the job requirements in your cover letter. For example, if the job involves oral presentations, communicate that this is a skill you have developed through your experiences.

Finding a job that you will enjoy can be a difficult process. Some people pay dues for years, while others seem to luck into good jobs quickly. Watching people over the years, we have seen that persistence pays off, and if you keep improving your application package, good things will happen for you.

In summary, science is a far more social endeavor than we had imagined when we started out. Effective communication with people around you can seem difficult and not relevant, but can be as important to your success as the science itself. Even though your main focus is figuring out how to do research effectively, it's well worth also investing time in developing the other skills you'll need to get the type of job you really want.

CHAPTER 8

Communicating
What You Find

Communicating is an essential part of doing field biology, although it requires very different skills than scientific investigation. Learning about nature is fun, but the field of ecology only advances when you communicate what you have learned. We have never been able to make a lick of sense of the argument that a tree that falls in the forest hasn't really made a sound if nobody is there to hear it. However, if you don't make other interested people aware of what you have learned, then from society's point of view, essentially nothing has been learned.

Not all attempts to communicate are successful, and this aspect of ecology has an enormous effect on whether your findings and ideas will have an impact. In the sixth and final edition of *The Origin of Species*, Charles Darwin (1889) included "an historical sketch of the progress of opinion on the origin of species." Essentially, Darwin explained why his ideas really were different from those of numerous predecessors who, by 1889, wanted some of the credit and fame for the theory that Darwin had expounded. Most of the authors were easy to deal with; they had simply missed the main points of the theory of natural selection.

However, one author was more troublesome for Darwin, and he wrote,

> In 1831 Mr. Patrick Matthew published his work on "Naval Timber and Arboriculture," in which he gives precisely the same view on the origin of species as that (presently to be alluded to) propounded by Mr. Wallace and myself in the "Linnean Journal," and as that enlarged in the present volume. Unfortunately the view was given by Mr. Matthew very briefly in scattered passages in an Appendix to a work on a different subject, so that it remained unnoticed....

Matthew understood the principles and their significance, but he didn't effectively communicate what he had grasped. He had the same impact as if he had never had the ideas in the first place.

This wasn't a one-time event. For example, MacArthur and Wilson (1963, 1967) revolutionized the field with their theory of island biogeography. Years before MacArthur and Wilson, Eugene Monroe proposed the same equilibrium theory, along with empirical support for the species-area relationship for butterflies in the West Indies, and detailed models to explain it (Monroe 1948 [his thesis], 1953 [an obscure proceedings]). Unfortunately, Monroe did little to communicate what he had found, and the scientific community remained unaware of his insights (Brown and Lomolino 1989). These examples illustrate that it matters where you publish. Make sure that you are reaching the largest and most appropriate audience. Both Matthew and Monroe are forgotten footnotes in the history of ecology because better communicators independently came to similar conclusions. How many Matthews and Monroes have there been, whose potentially revolutionary advances have never been repeated or communicated?

Publications are the currency of our field. Some journals are more influential than others and reach a much wider audience. Rating systems for journals have been developed to measure this influence. These ratings fluctuate and are available at several websites, including http://www.scimagojr.com/journalrank.php?category=2303. The best way to get a sense of the respected journals in your subdiscipline is to ask several experienced ecologists and to read (skim) a lot yourself.

Writing and talking about your work communicates your ideas and findings. In addition, the act of organizing your work by presenting talks and writing papers helps you figure out what you know, what you don't know, and how the various pieces fit together. Almost all seasoned ecologists will tell you that they often think that they have a pretty good grasp on a subject they are about to lecture or write about; however, once they sit down and look for the actual words they are going to use, they realize that they haven't thought through the ideas. The act of writing or speaking clarifies your thoughts and will probably be valuable for you, independent of the value of communication.

Many ecologists who communicate their work successfully use an outline to organize talks and papers. The outline can be either hierarchical, with roman numerals (Rick's preference), or informal, in a bulleted list (Mikaela's preference). When we're organizing our work, we use a list when we aren't sure about how to order the various ideas. Then we give each of the things we've written down a number or color code that helps to group the ideas that are similar or related. Next, we figure out which of these should go first and how to connect them to make a logical argument. If you think you don't like outlines, but you haven't actually used them to write a professional paper, we recommend you give them another try. It may seem like you're wasting time, but they make you more efficient in the long run and help you write a more organized paper.

If you took our advice from chapter 6, you will write up your results after each field season. The preliminary literature review that this requires will help give you a sense for how your work fits into the bigger picture. Writing up your results will also make it clear what you have nailed down conclusively and what parts of your argument are weak and need further testing. Coming to grips with what you have will also help you design the next steps. Doing this may seem like extra work. It's not. You wind up using a lot of this preliminary draft when you write up the paper for real. It also makes writing the final paper much less overwhelming. Finally, it is easier for colleagues or committee members to provide helpful feedback if they can read a manuscript rather than just listen to something vague about what you think you found. In the sections that follow, we offer some suggestions for organizing your work into (1) a journal article, (2) an oral presentation, (3) a poster, and (4) a grant proposal.

Journal Articles

Journal articles are the bread and butter of biologists. Writing papers can seem daunting at first, but as you begin to recognize the formula for writing them, they will become easier. Journal articles serve the important function of archiving what you have learned and making it available to the rest of the ecological community.

Expect to revise your manuscript. Rick sometimes likes to think of his first draft as a place to get his ideas organized and out of his head. He tries to approach his second draft much more critically, from a reader's perspective. Is the story clear? Does the logic follow? Does his writing express exactly what he was thinking? Sometimes he tries to imagine that his father is reading the manuscript. His father had no formal training and therefore didn't have the jargon and preconceptions of a trained ecologist. Rick asks: Would he follow what I am saying? How would I change my paper so that he would understand it?

Inexperienced writers sometimes imagine that they should sit down and write a polished version. Instead, it may be helpful to think of the writing process as four distinct steps (Lertzman 1995). The first step is figuring out what you have to say. Don't worry about grammar or organization during this stage—your goal is to get your thoughts on paper, as we described above. The second step is organizing your thoughts (with an outline or other technique that works for you) so that you can present a logical argument. The third step is putting down the words that make your argument. Read this draft critically, imagining how it will sound to your audience. Then, as a fourth step, carefully craft your writing so that it makes your points concisely and convincingly. Separating these tasks may help you get started. Including all four, if you don't already, should improve your writing and make the process less difficult. If your native language is not English, it may be a good idea to get a native speaker to help you with this last step.

Most journal articles are expected to follow a standard format: abstract, introduction, methods, results, discussion, and conclusion. (Even articles in *Science* and *Nature* are written in this format, although it is less easy to spot.) Most of your paper should be written in the past tense. You are describing what you did, what you found when you did it, and how you interpreted those findings. Some authors write the introduction in the present tense, describing the current state of knowledge.

In the following sections, we give you some information that will help you with each of the components of a journal article.

Title and Abstract

The paper starts with a title and an abstract (although we find it is easier to write these later, when we have a good sense of the main points and their significance). The title tells what the paper is going to be about. Don Strong, the editor of *Ecology*, says it should present the main result rather than just including the key words. For example, "Fire increases butterfly diversity in riparian and woodland habitats" is a more informative title than "A study of the effects of fire on butterfly diversity in two habitats."

The abstract provides a summary of the paper. It should include a couple of sentences each of rationale, main results, and interpretation. Throughout, be concrete; in other words, don't tell us that results are presented—tell us instead exactly what they are. The abstract has to be concise and clear. Far more people will read your abstract than other parts of your paper. Even if they do read the entire paper, reviewers and critics of all kinds will make their decisions about the paper and your story based largely on the title and abstract.

Introduction

Your introduction should present your question and explain why it is interesting. Your first paragraph or first few sentences should set the stage for your question. How do we (other ecologists) think about this subject? One effective way to begin your paper is by stating a problem or observation that everyone agrees is important and grabs our attention. If it is not obvious why your problem is important, then you must make a case for why we should care enough to read your paper. Could solving your problem shed light on a bigger issue, for instance? Think about what matters to your audience (e.g., conservation, basic physiology, theory) and frame your question so that they find it interesting.

We find an introduction that poses a big-picture question to be much more effective than one that starts with a description of an organism, study system, or topic (although the latter is how many students are tempted to begin). So for instance, don't tell us that you are interested in wooly bear caterpillars. Instead, lead with your general question: How do herbivores choose food? Then tell them how the study will add to this knowledge by considering the food choices of wooly bears. If your audience won't immediately find this question highly relevant to their interests, explain why choice of host plants is critical to understanding other ecological and evolutionary questions. In other words, make sure that you have explained why we as ecologists need the information you are providing. Don't just assert that it is critical to know the rates of parasitism of lemmings. Instead, explain why knowing about parasitism could help us understand why populations cycle. Wait until later in the introduction, or even the methods section, to tell us the natural history of your system. Beginning with a general "hook" will catch the attention of a larger audience.

You test the general question you led with by looking at a specific example. In your introduction, you may want to reference other research that informs your big-picture question, such as prior work on this question that has been conducted in other systems. But do not include references in your intro (or anywhere in your paper) just to show that you are familiar with the literature. You may want to explain why research in your system will contribute to answering the more general question.

We like to end our introduction by giving either a formal listing of the questions that we will answer (for a paper) or a brief answer to the question that we posed at the beginning (for a talk, see next section). What specific hypothesis or hypotheses are you testing? They don't have to be stated as null hypotheses (which have a bit more statistical precision but can be confusing to follow). These questions let the audience see where we are going to go in our methods and results. Also, because this formula is relatively standard, some readers will skip to the last paragraph of your introduction to decide whether the questions you plan to answer are interesting enough to keep them reading.

Methods

Methods sections often begin with a brief description of the natural history of the system. (This can also go near the end of the introduction, preceding the list of questions that the paper addresses.) Tell us enough, but only enough, natural history so that we can follow the important points of your experiments and interpretation.

The core of your methods section is the description of what you did to gather data. Your methods must be described clearly enough that the work could be repeated by someone else. This should include a description of where, when, and how you applied your treatments and took your measurements, as well as the statistical methods you applied. Make sure that the reader understands the motivation for each experimental procedure: instead of just launching into the details, start the description of each experiment with something on the order of "To test the hypothesis that wooly bear caterpillars choose the most common host plants, we did the following experiment."

Your methods section should be kept brief because methods are often boring to read and because reviewers and editors look unfavorably on papers with lengthy methods. Graduate students often assume that they need to explain every detail of their projects, but this is not true. Only include information that is directly relevant to the story that you want to tell. For example, perhaps you kept detailed data each day on temperature, percentage of sunlight, or precipitation because you thought that these might help explain variation in crawdad feeding events. However, you didn't find any such relationship, and your story evolved in other directions. Readers don't need to hear about these details even though you may be tempted to show them how thorough you have been. Information that isn't immediately relevant will only clutter up your presentation and obscure your main story.

Results

The results section tells what you have found—your data and statistical analyses. In general you will not be permitted to make a statement if you don't have statistical analyses to support it. Results should be described in a way that tells your story logically. Often students describe their results chronologically, in the order in which the experiments were conducted. But a topical organization that answers your main questions in a logical progression is generally more effective. We like to organize our methods and results sections by questions or experiments. We use subheadings as titles for each question; and we use these subheadings in the same order in both sections so that the reader can easily connect a method with the corresponding result.

Results should be reported in the text or summarized as figures or as tables (generally use only one of these three for any given result). Text is the default choice; if a result can be conveyed effectively by describing what you found, do so. Figures are particularly good at conveying relationships between factors, although actual values are generally obscured. Tables show exact values but are not good at presenting relationships between factors. Often the most informative way to present your results is to illustrate the data in a figure and then describe the effect size in the text. For example, if you are describing the effects of selenium on salmon fecundity, you might use a bar graph to show that female salmon produced approximately 60 eggs in low selenium streams and 20 eggs in high selenium streams. Then, in the text, "High selenium levels dramatically reduced salmon egg production. Female salmon produced only about one-third as many eggs in high-selenium streams as in low-selenium streams" (followed by a statistical test comparing these means).

Make your results, not your tables and figures, the star of the show. The tables and figures merely illustrate what you are describing in the text. For instance, "Adults consumed 40% more than juveniles (table 1)" is preferable to "Table 1 shows the consumption rates of adults and juveniles." Similarly, when citing published work, focus on the results, not the authors: "Males were bigger than females (Brown 2000)" rather than "Brown (2000) reported that the sexes differed in size."

Simple figures and tables are better than more complicated ones. Figures and tables should be described clearly with titles and captions, so that someone looking at them can make sense of the results without necessarily reading the whole paper. Your audience must be able to discern what you actually measured. This is often achieved by clearly labeling the axes of your figures. The fewer the number of treatments presented in one figure, the easier it is to grasp. Don't combine figures unless viewing all the information at one time adds meaning. If possible, identify the treatments in a legend rather than in the figure caption. Letters, numbers, and lines must be readable when they are reduced for publication, which means you often have to make them larger and thicker than the defaults on your software program.

The most commonly used figures are bar graphs and scatterplots. When you use bar graphs, it is easier for a reader to grasp the meaning if you use fewer bars. Under most circumstances, error bars should be presented with all bar graphs. These are important because they give the reader a sense of the noise around the signal. Usually you will use standard errors to show the noise or precision around your estimate of the mean. Standard deviations are used only when you want to show the amount of variation *per se*. Occasionally error bars make a figure so busy so that the signal becomes unrecognizable, and only under these circumstances should error bars be omitted. Scatterplot figures are also commonly used by ecologists. The line that describes the best fitting model can be added to the scatterplot when the model is found to be significant.

We find that diagrams, cartoons, and other graphic representations of ecological hypotheses are often informative, although they are less commonly used and generally belong in the introduction or discussion rather than in the results. These can be simple path diagrams that describe hypothesized causal schemes, or more complex figures that provide a visual representation of an ecological model. Many readers grasp concepts more readily through figures than through verbal or mathematical representations.

Tables should be used only when repetitive data are essential to tell your story. For many arguments, fewer data are more effective than more. Only include those variables that are relevant. Don't use tables (or any part of the results section) as a core dump for your field notebook. One common application for tables is to summarize statistical tests. For example, in an analysis of variance, the sums of squares, F ratio, degrees of freedom, and p-values all provide unique information. If this information is not required to make a convincing case, then include just the F ratio, degrees of freedom, and p-value in parentheses in the text.

To communicate effectively to a biological audience, your results should be presented in biology-speak rather than statistics-speak, and you should highlight the biological results and not the statistical tests. For instance, tell us "Males were twice the size of females (Student's $t = x$, df $= y$, $p = 0.0z$)" rather than "The Student's t-test with y degrees of freedom showed a statistically significant effect at the $0.0z$ level of gender on body size." Further, always present the effect size, not just the level of statistical significance— "Males were twice as big as females" rather than "Males were significantly bigger." The effect size tells us about the biological relevance of the result (see box 4 in chapter 4), whereas the statistical significance tells us how likely it is that the result was caused by chance.

As we mentioned in the section on methods, it may be tempting to include all of the experiments and observations that you performed. Don't do it. Include only those results that are connected logically—that tell one coherent story. Variables and effects that are not relevant to the story should be omitted, or the audience will be distracted from the main points. Many authors make the mistake of trying to include all the data they have instead of thinking about what pieces are needed to tell the best single story. If you feel compelled to include data for archival purposes, stick them in an appendix or online supplement rather than bloating your results.

Discussion

Use this section to explain what your work means. To do this, it is often a good idea to restate concisely the most important result before you interpret it. How do you make sense of what you found? Do your results resolve the question that you posed in the introduction? What evidence have other studies brought to bear on the question? Then as they become relevant to your story, add in the other results of your study and interpret them. Often the results of experiments will suggest subsequent hypotheses, which can be integrated into your discussion. You may be able to generalize from your results in conjunction with those of others. Do any useful models or paradigms result from this work?

Believe your results and interpret them as such. If you didn't find what you were expecting, don't excuse your results and talk about how they might have been different with a larger sample size, or if you had controlled for other factors, or if you had done the work in a different place. If you don't believe your results yourself, don't take our time telling us about them. If the evidence supports your hypothesis but you still don't believe your results because you feel unconfident and doubtful about everything, talk about this with a therapist, but don't let these doubts pervade your presentation. Also remember that most discoveries are surprises. If you already know the answer, then the question isn't particularly interesting.

Throughout your paper you should tell a cohesive story. Don't wander from your central point. Rather, your writing should present a tightly reasoned argument that is evident from start to finish.

Conclusion

Papers should end with conclusions (although these are often missing). The conclusion, like the abstract, is a concise summary of your results and their significance. End with a sentence or two that states the important consequences of the findings. Leave us with the take-home message—and don't have too many take-home messages. Most papers have only a single real lesson. Make sure that it doesn't get lost but instead is painfully obvious for those people who will read only the conclusion.

It may be useful to once again state and answer the question that you posed at the start. Don't trail off with some weak non-conclusion like "this is a good system" or "more work should be done." Of course more work should be done following every study. Instead, leave us with what you have learned. If we remember anything from this work, what should it be?

Box 6 provides a summary and checklist of our suggestions for writing journal articles.

The publication process can be emotionally brutal and requires a thick skin. All ecologists get their manuscripts rejected. Ecologists who succeed at publishing the most also experience the greatest number of rejections (Cassey and Blackburn 2004). Even established professors experience a rejection rate of 22%. In a careful statistical analysis across multiple journals in ecology, papers that were rejected and resubmitted were ultimately more often cited than those that were published without resubmission (Calcagno et al. 2012). Papers published without resubmission probably don't contain any ground-breaking ideas or data and were assimilated without question.

Although it's painful, the review process does make your papers better. Reviews from journals indicate how two or three readers perceived your paper. If they missed important points, other readers are likely to miss the same points. Take the comments of reviewers seriously; they almost always have useful ideas for improving the manuscript. When a paper gets rejected, put it aside for a day or two, and then make changes that will address the concerns of the reviewers whenever possible. If you are being given the opportunity to resubmit your manuscript, address every point that the editor and reviewers raised both in your cover letter and in the text. Letters from editors generally

Box 6. *Journal article checklist*

Title

☐ Does the title summarize the main result?

Abstract

☐ Does the abstract tell your story, very concisely?

Introduction

☐ Does the beginning of your introduction "hook" the reader by setting the stage for the question(s) your paper answers?

☐ Do you explain and justify your question(s) instead of just extolling the virtues of your study organism?

☐ Do you briefly summarize previous work that informs your current question(s)?

☐ Do you end your introduction by clearly listing the question(s) your manuscript addresses?

Methods

☐ Do you briefly explain the relevant natural history of your organisms or study system, if you haven't already included it in your introduction?

☐ Do you describe your methods thoroughly enough that another ecologist could repeat your experiment, but briefly enough that space-pressured journals won't send your manuscript back?

☐ Do you start the description of each experimental method with a phrase justifying why it was done?

☐ Do you include a brief explanation of each statistical procedure you used?

Box 6. Continued

☐ Do you include only the methods relevant to your overall story?

Results

☐ Are your results presented in a logical order to help your reader follow your story (not in the order in which you did your experiments, if that is different)?

☐ Have each of your results been presented once, and only once (in the text, a figure, or a table)?

☐ Does your text inform your readers of your results as much as possible, instead of simply referring them to your figures or tables?

☐ Do you describe your results in biological rather than statistical terms?

☐ Do you present effect sizes for each of your results?

☐ Do you present each of your results in terms of your overall story?

Discussion

☐ Do you restate your main results very briefly and interpret them?

☐ Do you generalize to larger ecological concepts where appropriate?

☐ Does the information in your discussion relate to your initial question(s)? Does your story seem cohesive?

Conclusion

☐ Do you hit your reader over the head one last time with your take-home message?

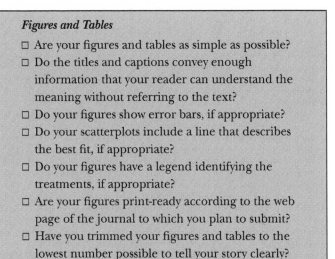

Figures and Tables

☐ Are your figures and tables as simple as possible?

☐ Do the titles and captions convey enough information that your reader can understand the meaning without referring to the text?

☐ Do your figures show error bars, if appropriate?

☐ Do your scatterplots include a line that describes the best fit, if appropriate?

☐ Do your figures have a legend identifying the treatments, if appropriate?

☐ Are your figures print-ready according to the web page of the journal to which you plan to submit?

☐ Have you trimmed your figures and tables to the lowest number possible to tell your story clearly?

English as a Second Language

☐ If you are not a native English speaker, have you gotten someone confident with English to review your writing?

sound more negative to the uninitiated than they were intended to. Few manuscripts are actually accepted on their first submission. Having your submitted paper 'rejected with the opportunity to resubmit' is the new 'accepted' (if you can address the reviewers' concerns). As a grad student, you can learn a lot about the review process by getting involved yourself, either formally or informally. Offer to review papers for other grad students, for your major professor, or for journals if your professors are handling editors.

Oral Presentations

Hearing a talk is a very different experience from reading a journal article. As a speaker, you should be aware of the differences so that you can use them to your advantage. Interacting with a person is far more compelling than reading a book. Think about how likely you are to put down a book without finishing it compared to how likely you are to walk out of a play or movie.

General Organization and Presentation

The more you involve your audience, the more successful you will be at holding their attention and having them remember what you say. As a result, you definitely do not want to read your talk; it's far more difficult to listen to someone who is reading aloud than to someone who is speaking. A conversational tone is easier to absorb than a speech. If possible, you should give your talk without notes, but this is less critical than not reading it. If you are worried about forgetting your talk, use an outline or let your slides lead you. If there are specific essential concepts or points that you are prone to forget, link them to a particular slide. When you get to that slide (usually an image rather than text), it is your cue to remember to provide a particular piece of your story.

If you are not a native speaker of the language you are presenting in, giving a talk can be even more daunting. If absolutely necessary, read your talk, but practice making it sound as conversational as possible to keep your audience's attention. Practice pronouncing key terms until you have made them as clear as you can. It also helps to put those words into your slides in case people still don't catch them. It can be tempting to speak quietly or very quickly, but resist the urge—your audience will enjoy your talk more if you speak loudly and enunciate.

When you are giving a talk, look at the people in the audience; eye contact will help to involve them. A common mistake is facing the screen and talking to your slides; instead, talk to your audience. If you must see your slides, glance at your computer rather than turning your back on the audience. Leave the room as bright as you can so that you can see the faces of the audience and they can see yours. It is a biological fact that dark rooms and slides with dark backgrounds put people to sleep; don't use them. A lot of light is much more important than having photographs that show up really well on the screen. Which would you prefer, having some large fraction of the audience dozing while really sharp pictures go up on the screen or having the audience alert and attentive at the expense of some photographic quality?

Stay close to your audience. This lets you relate to them more effectively. If the podium is too far away, move it closer or don't use it. Come out from behind it and address the audience directly. Walk around a bit in the space you are given as well. It is amazing how much walking from one side of the screen across to the other helps keep your audience alert and focused. Speak to the audience in the corner farthest from where you are standing—it will help you remember to project your attention throughout the room instead of just to the first few rows.

Tailor your talk to your particular audience and keep the information clear. Tell people things that they have the background to relate to. Imagine what the interests of your audience are likely to be and prepare your talk with that bias in mind. Avoid jargon (abbreviations, specialized words, measures, techniques) that is not familiar to most of your audience. Once people tune out or get lost, it is hard to get them back. You cannot expect people to grasp equations or complicated theory that they have not come to grips with previously. When readers come upon new or complicated material in an article, they can slow down, digest it, and reread it until they understand it. There is little chance for this to happen in a talk, so don't lose your audience by including this kind of material. It is fine to include a theoretical motivation for your work or to develop a new theoretical argument, but use words rather than equations if you are presenting a talk, and be especially careful that you are clear and the material is accessible. If you are presenting a talk that is mostly theory, you may have no choice but to include an equation or two. Spend a lot of time with any equation, explaining in words and perhaps simple graphs what each of the clusters of terms means to a biologist. It may help to circle relevant terms in the equation and then explain in plain English what they mean.

Structure

Good talks require as much structure and planning as good papers. Don't imagine that you can just wing your talk off the top. When you are organizing your talk, build it around your take-home message. Figure out what your punch line is, set it up right from the start, tell the audience what it is going to be, deliver it, and then remind the audience what it was during the conclusion. Someone, supposedly either Aristotle, Dale Carnegie, or Winston Churchill, came up with a pithy way to remember this strategy: "Tell them what you're going to tell them, tell them, and then tell them what you just told them." Those listening to your talk should be able to grasp your take-home message even if their thoughts have drifted away at some point or they came into the room after the talk began.

A paper requires elaborate documentation of the existing literature, of your detailed methods, of the statistical analyses, and so on. These are generally boring to listen to and should be minimized in a talk (though you should be ready to talk about them if anyone asks about them at the end). Your talk must have one, single coherent story to tell. Don't be tempted to include all the loose ends—just your best single story. This is true for a paper but even more so for a talk. Stress the concepts and not the details. Try not to switch gears and tell two stories during your talk; your audience won't take much away from such a mixture. If you feel that you don't have enough material for one coherent story, then spend some time thinking about how to organize it so that it seems like a single story, and then more time smoothing the transition between the pieces. One way to do this is to present a single question (or short series of interrelated questions) in the introduction and one punch line that integrates all of the parts of your story in the conclusion.

We like to use signposts to let the audience see the structure of the talk and how each piece fits into the bigger picture. When you're reading a paper, you rely on the sub-headings and paragraph indentations to recognize transitions. The audience listening to a talk needs similar signposts. Rick often likes to make an outline slide of his talk or have it up on the whiteboard if one is available. Then he refers to the outline at various times during the talk. It helps people follow his train of thought if they can see where he is going throughout the presentation. (Remember, though, that the whiteboard will not be visible if the light is dim.) Mikaela uses a higher tech approach. She puts up an outline slide early in her talk and repeats it at various times throughout her talk, highlighting where she is in the outline at each point.

Let the questions dictate the outline. Don't use the traditional methods, results, and discussion that you would in a paper. Instead, integrate these three for each question or sub-question that you address. For each experiment, tell the audience in a sentence the question that was asked. Describe the approach you used in a sentence or two, and then describe the result. What does this result mean? If it stimulated the next question, explain how. Then repeat this sequence for the next experiment. Make sure that the audience follows your logical path.

Methods

The description of your methods should be very abbreviated and contain only what is absolutely essential for you to tell your story. A talk should not attempt to provide the listener with the ability to repeat the experiment; unfortunately, many speakers make the mistake of including far more methods than are necessary or interesting. If you need to include a method in your talk, illustrate it with a photo rather than text. Very abbreviated methods are often effectively integrated with results in talks.

Results

When presenting your results, aim for simplicity. Simplicity is important for a paper, but critical for a talk. Pictures are better than words or tables. Make sure that you explain the axes of graphs. The audience isn't already familiar with the variables you put on each axis and needs to identify these before the data you are presenting can be appreciated. Keep the graphics simple. Three dimensions and a bunch of colors rarely help tell the story. In general, tables are much less effective in talks than in papers and should be used sparingly, if at all. Any tables you do use must be simple, with large, easily read characters.

Similarly, each slide should make only a single point. Never put more on a slide than can be presented easily. Don't put up slides with text that you merely read. Text slides should be as abbreviated as possible—a few words or phrases. We don't like full sentences. Put no more than ten or fifteen words up at a time, and use a font size of at least 24.

Let your slides guide your story, but don't make them your focus. Plan to give your talk as if you didn't have slides. (In case there is a technical failure, you should always be able to give your talk without using slides at all.) Slides should be the background that illustrates what you are saying. Don't structure your talk so that it is a progression of

"This slide shows this and this next slide shows that." You'll give a better talk if your slides illustrate the story rather than if they are the whole story. Instead of reading the exact words from a slide, share the spontaneity of putting the words together to express your thoughts. Focus on connecting with your audience, not with your slides. If you must use a pointer, a physical pointer, your arm, or a meter stick can be better than a laser pointer; if for some un-known reason you feel obliged to use a laser pointer, don't wiggle it about, circle the object repeatedly, or let it wan-der annoyingly across the screen.

Preparing Your Talk

Spend extra time preparing your introduction and con-clusion. These are the only parts that some of the audi-ence will hear. Everyone is most alert at the start, so tell them what questions you are going to ask and what you found. Even if you lose some of them at some point, they will have heard your punch line. Similarly, if they haven't followed all of the talk, the conclusion should provide the take-home message. Students are often impressed by sev-eral established scientists at UC Davis who nod off conspic-uously during talks but always seem to ask good questions afterwards. When this occurs, a lot of the credit should go to the speaker for highlighting the important points be-fore the lights went out and then again after they came back on.

Practice your talk before you give it for real. The more you practice it (especially in front of an audience), the bet-ter it will be and the more confident you will feel when it's time to give it for real. If a rehearsal audience is not available, saying the actual words out loud is much better than thinking them. Mikaela used to think it was too embarrassing to practice a talk out loud with her housemates overhearing her. Then she learned it was more embarrassing to give a poorly practiced talk. Recording a practice (or actual) talk will help you learn your talk quickly and will also help you improve it if you give it repeatedly. If Rick has to give a talk that he really cares about or if he is strapped for time, he records a practice and then listens to it (or at least has the recording on) several times while he's doing other things. It is amazing how much this helps. If you're really brave, videotape your talk and review it to learn what aspects need improvement.

Editing Your Talk and Getting Ready

Show only as many slides as your audience can digest. As a rule of thumb, use about one slide per minute of talk. Many speakers make the mistake of including far too many slides and then needing to rush at the end or annoy the audience by going overtime. To avoid this, time yourself each time you practice.

We like to pause after we have made an important point. This provides emphasis and gives listeners a chance to make sure they absorb the last message. When you first begin giving oral presentations, your nervousness is likely to cause you to speak faster than you practiced. Practice making well-timed pauses to help you to pace your talk. Stopping to take a deep, slow breath helps your audience digest what you just said and can help you stay focused.

Show Time

A talk from a person who is slightly nervous is often better than a talk from someone who lacks sufficient adrenaline. However, excessive nerves can make a talk difficult to follow. Convert your nervous energy into larger, more flowing motions rather than small, repetitive jitters. Remind yourself that the more times you give talks, the easier they become. A mistake many beginning speakers make is to be self-deprecating and apologetic. Replace this with enthusiasm; your feelings become contagious.

One of the best parts of a talk is the question session at the end, although this can be the most intimidating. We like to leave a lot of time at the end for questions (10–15 minutes for an hour-long talk and 2–5 minutes for one that's 12–15 minutes). We already know what we have to say, but we're excited to hear the spins that other people will place on our results. Often new and exciting ideas come up in the questions after a talk. We sometimes ask a friend to take notes during the questions so that we don't have to remember all the suggestions. If the room is large or the questioner soft-spoken, repeat the question for the audience before answering it. Make sure you understand a question before responding to it. It is fine to paraphrase the question and ask the questioner if you have it right. If you don't know the answer to a question, it's fine to say so. If you think the point raised is an interesting one, you can say you will think more about it and design an experiment in the future to test it. You might ask if the questioner can think of a way to test the notion that he or she is bringing up. Especially if the talk you are giving is your oral exam, try not to be defensive when people question you about your talk. Students who fail oral exams are generally not ones who are the least prepared but rather the ones who become defensive and argue with their committees. If an audience member is aggressive about asking questions and won't give up the floor, you can say that you would like to move on but you would be happy to talk more after the session is over. By the way, remember not to be that person when you are in the audience.

Box 7 presents a summary and checklist of our suggestions about talks.

Posters

Posters have become the most common medium at some academic conferences. They should be much more like talks than papers in their structure. However, most posters suffer from being prepared like manuscripts. Remember that people at meetings are burnt out. Do you enjoy reading a lot of fine print when you are viewing posters? We don't. Instead, we want the take-home message in a simple, readily available package. Posters are to scientific communication what *USA Today* is to journalism. You should present only the headlines and the briefest explanation to make your point. Your poster should present a short summary of your take-home message and should encourage conversation if you happen to be present. Everyone who walks by your poster should immediately know your question and

Box 7. *Oral presentation checklist*

See also box 6, "Journal article checklist," for reminders of good communication habits in ecology.

General Structure and Presentation

☐ Are you able to give your talk (from the slides or an outline) without reading it?

☐ Have you practiced making eye contact with your audience (instead of with your slides), moving about the room enough to keep your audience engaged, etc.?

☐ Have you carefully examined your talk for jargon you might not even realize you are using?

☐ If your talk includes an equation, have you planned how you will make it readily accessible to your audience?

☐ Do you present your information as one single, coherent story to help your audience follow you?

☐ Does your talk includes signposts so that your audience follows the structure you have created?

☐ Do you use a large font size (24 point or larger) and include very few words (10–15 maximum) at a time per slide?

Introduction

☐ Do you structure your introduction around your take-home message?

☐ Do you eliminate most of the citations and other details you would include in a manuscript to help keep your audience's attention focused?

☐ Do you end your introduction by showing an outline slide that clearly indicates the questions you will address?

Methods, Results, and Discussion

☐ Have you integrated your methods, results, and discussion in a way that makes your story easy to follow (even if that means doing a separate series of methods, results, and discussion for each question)?

☐ Have you minimized (that is, practically eliminated) your methods?

☐ Do you explain how the results of your first experiment generated your next question and experiment, so that your audience understands the relationship between the parts as the story develops?

Conclusion

☐ Do you leave your audience with one take-home message?

Figures and Tables

☐ Do you show your results in pictures and figures instead of just describing them?

☐ Do you minimize or eliminate the use of tables, since they are hard to grasp during a talk?

☐ When presenting your figures, have you planned to indicate to your audience what the x and y axes are?

☐ Are your graphics simple, so that each makes only one single point?

Preparing your Talk

☐ Have you practiced your talk (especially the introduction and conclusion) until you are absolutely comfortable with the information in it?

Box 7. Continued

☐ Have you created roughly one slide for each
minute of your talk?
☐ Have you timed yourself to make sure your talk
does not go overtime?
☐ Are you prepared to give your talk without any
slides at all in case of a technical problem?

the answer. Those people who are interested can ask you
to explain in more detail and can read the paper when it
comes out.

The title is the only thing that many people will see. It
should summarize your results in one phrase. Like the title
of a paper or talk, it should give the main message rather
than a list of the characters or the question. For example,
"Interactions between student protestors and campus au-
thorities" is less effective than "Chancellor defends pepper
spraying of peacefully protesting students." Since most
conferences have many posters, you need to compete for
an audience. Your title and the visual layout of your poster
must be compelling. A poster is analogous to an elevator
speech, in which you have ten seconds or so to sell your
work or convince someone that what you do is valuable
(Erren and Bourne 2007).

Structure

Your introduction should be no more than a few sentences stating the conventional wisdom or explaining the

justification for your question. Next, present your results as pictures (figures and photographs) that tell your story. Move logically from one result to the next, making sure not to include more information than your viewer can easily and quickly digest. You should either skip the methods completely or include only enough to make your results meaningful. The details of your experimental design, sample sizes, and so on should not be included. After each result, you can include one sentence of "discussion" that makes each result more general or relates it to your big question. At the end of your poster you might want to include a sentence or two that explicitly answers the question that you posed at the start. Another useful way to conclude is with a sentence or two (but no more) explaining the significance of your results and how they fit into the big picture. This can be labeled as your conclusion. Usually the conclusion will be the section of your poster that will be most noticed after the title and abstract.

After you have worked out the pithy content of your poster, spend some time figuring out how to present it so that exhausted conference attendees can follow you. Again, images can illustrate your points and help you cut down on text. Use large or color type to draw attention to your organization and main points. As a rule of thumb, sans-serif fonts are better for titles and headings and serif fonts are better for full sentences; it's fine to mix both font types in a single poster. Keeping a lot of white space can help focus the viewer's gaze. Variation in font sizes and colors can help the viewer grasp your organizational structure. Finally, get feedback before printing it.

One advantage of presenting a poster is that you can "walk" interested people through your story. This is more effective than asking them to read the thing. In addition, they have the opportunity to ask you questions about things they don't understand or suggest other experiments and directions. As such, it seems like a good use of your time at the conference to hang out with your poster and interact with viewers as much as you can. Practice explaining the content of your poster beforehand so that you are ready when someone stops to ask about it.

It is often helpful to include a photo of yourself and co-authors on your poster so that interested people can find you during the meeting. Contact information including your address and email should also be included. Some presenters like to have handouts of the poster on 8.5 × 11-inch or A4 paper, or copies of related journal articles.

The poster that we have described contains less than one tenth the number of words of most posters at ecology meetings. It tells only a single story and does this using only headlines. It contains no or few references, and no methodological details. It has figures and photographs but rarely tables. Details and statistical analyses are not included. It is much more effective at conveying information than the poster that is essentially a manuscript pasted on a board.

Box 8 presents a summary and checklist of our suggestions about posters.

Box 8. *Poster checklist*

Refer also to box 6, "Journal article checklist," for reminders of good communication habits in ecology. Because your poster will rely much more on pictures and figures than on words, you may especially want to refer to the section labeled "Figures and Tables" in box 6.

Title

☐ Does the title summarize the main result?

Introduction

☐ Do you limit the introduction of your question(s) to one or two sentences?

☐ Do you clearly present the question(s) your poster will answer?

Methods

☐ Is your methods section extremely brief?

Results and Discussion

☐ Are your results presented mainly as graphics (bar graphs, scatterplots, etc.)? Do you show the differences in treatments with photographs where appropriate?

☐ Do you briefly explain the significance of each result?

☐ Do you present each of your results in terms of your overall story?

Conclusion

☐ Do you include a sentence or two briefly answering the question(s) you posed at the start?

> *Box 8. Continued*
>
> **General**
> ☐ Does your poster contain only the headlines?
> ☐ Does your poster use variation in font sizes, font styles, and colors to help the viewer grasp your organization?

Grant and Research Proposals: Selling Your Research Ideas

The purpose of grants and research proposals is to sell your plans about work you want to do. You want your committee to agree to give you a degree if you fulfill the objectives in your research proposal, and you want people to give you money in response to your grant proposal. In addition, your proposal provides two less obvious functions; it forces you to develop a research plan, and it forces other people to consider your ideas more carefully than they might otherwise so they can give you better feedback. Nonetheless, grant and research proposals involve more salesmanship than research talks or papers. Therefore, a slightly different emphasis is required. As you prepare a proposal, focus on three things: (1) novelty and justification, (2) clarity, and (3) feasibility.

Your proposal outlines what you want to do. First, it must be exciting and original. You must convince the reader that your work will forward your subdiscipline or the way in which people apply science to solve problems. Obviously not every proposal is going to change the way all scientists think, but those people who work on your question or on your system should be influenced by your work. If it is not clear to you how this will happen, think hard about how to justify your work in these terms. Emphasize this justification throughout your proposal. If justifying your proposal sounds too vague, think about answering questions such as: What makes your proposed work significant? What is the value of the work? How might other people use your results? How will other people inside and outside of your field view the contribution of your work? The biggest mistake that students make when writing proposals is not including enough justification.

Second, your proposal must be simple and clear, even more so than a scientific paper. Reviewers often get many proposals to read at one time, and all reviewers have better things to do than read them. Unlike scientific papers, these proposals may not be about subjects that the reviewers are already interested in or knowledgeable about. From the reviewer's first glance at your proposal, you have only a few seconds to convince him or her to pay attention and read further. Then you have only a few minutes to convince the reviewer that your proposal is worthy of funding from a budget that can in many cases fund fewer than 10% of the proposals in the stack. If your writing is not clear and concise, the reviewer may not do the work required to figure out what you are trying to say. The proposal must be convincing to both the meticulous reader and the rapid skimmer. A well-known colleague who serves on many NSF panels calls this the "two glasses of wine problem." He does all of his reviewing at the end of the day after two glasses of wine at dinner. Successful proposals must be clear enough to make sense to him under those conditions.

Third, you must convince readers that your proposal is feasible. Nobody is going to give you money or assurances of a degree unless they are convinced that you can complete the work you propose and that your work will answer the interesting questions that you have posed. There is an inherent contradiction in this process, since your proposal must appear both feasible and novel. You must convince people simultaneously that your ideas are important and ground-breaking, and that the experiments you propose can be accomplished. The best way to convince people that you can pull off your experiments is to use techniques that have been used before and include citations for those techniques. It is even better to be able to say that the techniques are old hat for you or people in your lab. An excellent way to show that your plan is feasible is to present preliminary data. For a research proposal this often involves doing a first year of fieldwork that addresses the main question. This heavy emphasis on preliminary results frequently means that researchers propose work that they have largely completed and use the money to generate the next set of preliminary data.

Organizing Your Proposal

The organization of a proposal differs slightly from that used for talks and papers and is generally less canalized into a specific style. Different universities and granting agencies require different organization, and it is important to know these requirements and to fulfill them. Below we consider the form and content of a proposal that would apply generally for many graduate groups and funding agencies. Many proposals include (1) an abstract or project summary, (2) an introduction, (3) explicit objectives, (4) experiments, justification, and interpretation that correspond to each objective, and (5) a budget. Other sections that are often helpful include separate discussions of the significance of the experiments, the potential pitfalls associated with the experiments and your solutions, a timetable for completion of each experiment, and a justification of the budget. You can get more detailed advice about preparing proposals in a recent book by Friedland and Folt (2009).

The abstract or project summary shares many similarities with those for scientific papers. It comes first, though we write it last. It needs to be crystal clear and capture the excitement and rigor of the proposal. In it, describe the big-picture problem that you are addressing. Next, emphasize a justification for your work and explain its significance. Describe what you expect to find and explain why your results will be influential in your field. The project summary generally presents fewer results than an abstract for a paper, but can contain a few sentences about your approach.

The introduction to the proposal must get the reviewer excited about your questions. Justify why this work is important. This is difficult. Even after giving this advice to grad students, we find that more often than not their proposals could use more justification. How does your work relate to the big questions in ecology and why should we care? Frame your work in terms of the questions that you will address rather than systems that you will use. This advice holds even if you chose your project because you were interested in the system. In fact, it holds especially if you chose your project because of your system. Rather than saying "This question is transformational," explain why it is transformational. For example, if you are studying chytrid fungus population dynamics, say, "Chytrid fungal infections threaten amphibians worldwide. To better conserve amphibians, we must understand the factors that affect fungal transmission and population dynamics," and so on.

It is often best to first present the general question and then describe how your specific research on a particular system will address that broader question. Start general and get specific. Your introduction should explain what has been done to date to motivate your question. Here you can also tell us about the natural history of your system, but only if this information is immediately useful in understanding how you will answer your question.

Next present your objectives. These can be long term (more than you can accomplish now to address a big-picture question) and short term (the actual goals of experiments in this proposal). The objectives should be presented explicitly and should be numbered. With each objective, you can include the hypothesis that is tested and a rationale for that objective. Using Christopher Columbus as an example, van Kammen (1987) differentiates objectives, justification, and hypotheses. If Columbus submitted a proposal to Queen Isabella and King Ferdinand, his *objective* would be to establish a new trade route to India to bring back three ships full of spices. He would *justify* his proposal by explaining that a water route to the west could be faster and less expensive than currently used routes and that such a route would increase their wealth and international power. By fulfilling these objectives he would test the scientific *hypothesis* that the earth was round. He would further justify his proposal by attempting to convince Isabella and Ferdinand that it was feasible to accomplish these objectives and that he, Columbus, had the necessary know-how and experience.

Each objective should be addressed by specific experiments. It is often useful to number these experiments exactly as you have numbered the objectives. A rationale and an experimental design should be presented for each experiment. Describe how you would do each experiment with sample sizes included, and demonstrate that you can accomplish each procedure by presenting preliminary results or citing similar methods. Finally, describe how you plan to analyze the data from each experiment.

Tell us how your results will be interpreted: "If experiment 1 gives this result, I will conclude the following." Interpretation of the results may or may not be its own section in the proposal. Remember that although hypotheses in ecology must be testable, they are not necessarily falsifiable or mutually exclusive. At this point you might want to include another section or paragraph entitled "Significance" if the importance of your work has not been extensively discussed and stressed.

We like to include a small section detailing potential pitfalls. This section provides damage control. Anticipate questions that the reviewers are likely to have and address them here. Try to explain how you will turn apparent misfortune into a situation in which the field will learn a lot. Describe here how you will interpret experimental outcomes that differ from those you anticipate. The best projects are those that give interesting results no matter what the outcome. If you have designed a research program that will let you gain new and useful perspectives about nature no matter what the outcome, make certain that you stress this feature.

We also like to include a timetable for our objectives and experiments. This helps establish that we have thought about how and when we will get everything done. A timetable helps make the work appear feasible and is useful to refer to when doing the work.

If you are applying for money, include a realistic budget that will enable you to complete your project. This is an itemized list of your expected costs. Explain why you need each piece of equipment, supplies, field assistants, travel money, and so on.

Box 9 summarizes our suggestions about grants and research proposals.

Three things about the granting process should be kept in mind. First, grants are competitive, and it often takes several attempts before a grant gets funded. Don't get discouraged. At the same time, take the comments to heart. We find it helps us get our emotions under control if we put the comments down for a few days after getting painful criticism. It can also be frustrating to get comments that seem to miss the point. If a reviewer missed our point, it indicates that we need to rewrite the proposal so that two

Box 9. *Grant and research proposal checklist*

Refer also to box 6, "Journal article checklist," for reminders of good communication habits in ecology.

General

☐ Is your proposal novel and exciting, and have you explained why to your reader?

☐ Do you explain the value to the larger ecological community of your proposed work?

☐ Is your proposal simple and clear, easy enough for an exhausted non-ecologist to understand at the end of a long day?

☐ Is your proposal feasible, and have you explained this in a way that will be convincing to your readers? When possible, have you proposed to use established techniques and presented preliminary results?

Project Summary/Abstract

☐ Does your project summary capture the excitement of your proposed research?

Introduction

☐ Do you take extreme pains to justify your proposed work?

Objectives

☐ Do you state each of your objectives explicitly?

☐ Do you justify each of your objectives?

☐ Do you articulate hypotheses that address your objectives?

☐ Have you designed and described experiments that address your hypotheses and objectives?

Box 9. *Continued*

Interpretation, Significance, and Budget

☐ Do you describe how you will analyze your findings and evaluate each hypothesis?

☐ Do you highlight the significance of your potential findings?

☐ Do you include a budget, if appropriate?

☐ Are you truly excited to do the work if it gets approved?

glasses of wine won't keep the next reviewer from following our logic. Almost invariably, the comments will contain extremely useful suggestions as well as some misconceptions. If you are resubmitting a proposal, make sure to address all of the comments that you received from reviewers.

Finally, don't let the granting process dictate what your questions are. Of course we like funded grants for the way they signify the approval of our peers. In addition, some projects require money to pursue. But there are plenty of ecological questions you can address for relatively little money. The granting process is a very conservative one; only ideas everyone is already comfortable with get funded. This retards innovation. We have seen over and over again how a few dollars can get graduate students and senior faculty alike to change their research priorities and pursue projects that were not necessarily burning questions for them. Our advice is to follow your own intuition. Proposals are approved by a committee. Why give up something that is as personally important to you as your research direction to an anonymous committee? Would you let a committee approve your choice of a partner over the next three to five years?

Hard work often determines productivity, and productivity often determines success. Pick the questions that are most exciting to you whether you get funding or not and you are more likely to work hard enough to be successful.

CHAPTER 9

Conclusions

There is a card game called Mao that is popular on several university campuses. One of the rules of Mao is that players cannot ask or explain the rules. New players joining the game must figure out the rules by observation and trial and error. A player who fails to follow a rule is given a penalty. Doing field biology can be a lot like playing Mao. The rules of field biology, and academia more generally, often go unstated. In this handbook we have attempted to make the unstated basic rules of the game explicit. You may or may not wish to follow the rules. But you might as well know what they are because you will face the consequences for choosing not to follow them. Unfortunately, life and ecology are both complicated, and there are *also* long-term consequences for following the rules too closely. Below we highlight some of the rules of our game as well as some of the potential costs of following them too assiduously:

Rule 1. Manipulative experiments are a powerful and highly respected technique to establish cause-and-effect relationships in ecology. Experiments lend credibility to your study.

Cost of Rule 1. Experiments are only as good as the intuition that went into the hypotheses being tested. Make sure you find time to know your organisms or else your experiments won't teach you much. In other words, make time for natural history observations and surveys.

Rule 2. Test clear hypotheses by using inferential statistics.

Cost of Rule 2. Don't get drawn into treating ecology as a science of truly falsifiable hypotheses and universal laws. Generate alternative hypotheses and evaluate the relative importance of each one.

Rule 3. Increase the statistical power of your experiments with a large sample size of randomly assigned, independent replicates.

Cost of Rule 3. Replication comes at the expense of scale and therefore of realism.

Rule 4. Plan your observations and experiments carefully and evaluate your results frequently.

Cost of Rule 4. Don't get trapped insisting on answering your initial question. Keep working even when your questions and tests aren't perfect. Be opportunistic and pay attention to the directions that your system is trying to send you.

Rule 5. Write proposals and apply for funding.

Cost of Rule 5. Unless you like administration, don't let writing proposals replace fieldwork for you. The granting process is very conservative, so do the projects that are the most exciting to you even if you don't get funded.

Rule 6. The currency for researchers (and grad students) is publications. If you're a grad student who still holds an undergrad mentality that grades and classes are useful currencies, realize that the rules have changed.

> *Cost of Rule 6.* Just as grades never did perfectly reflect what you learned from classes, a long list of grants and publications does not perfectly reflect learning about nature and advancing the field.

Unfortunately, the rules reward short-term goals that may not be consistent with your longer-term goals. The good news is that most of us got into this business because we like being outdoors and learning about nature. You can and should make your job reflect your interests. You are likely to have more control of this as your career advances. While you play the game, keep your eye on the big prize: your own personal and professional priorities. This is your life! You will be more successful if you're enjoying it.

Acknowledgments

We have gathered the advice of our teachers, role models, and colleagues together in this handbook along with our own personal experiences. Many people have shaped how we go about doing ecology, and we have borrowed heavily from what we have been taught formally and informally. We thank Anurag Agrawal, Winnie Anderson, Jim Archie, Shelley Berc, Leon Blaustein, Gideon Bradburd, Liz Constable, Will Davis, Teresa Dillinger, Hugh Dingle, Alejandro Fogel, Jeff Granett, Patrick Grof-Tisza, Jessica Gurevitch, Marcel Holyoak, Henry Horn, David Hougen-Eitzman, Apryl Huntzinger, Dan Janzen, Sharon Lawler, Rich Levine, Monte Lloyd, Greg Loeb, John Maron, Rob Page, Sanjay Pyare, Jim Quinn, Dave Reznick, Kevin Rice, Bob Ricklefs, Tom Scott, Jonathan Shurin, Andy Sih, Chris Simon, Dean Keith Simonton, Sharon Strauss, Don Strong, Jennifer Thaler, Will Wetzel, Neil Willets, and Louie Yang, all of whom made valuable contributions to this book. We thank Wang Dehua for translating the first edition into Mandarin. We also thank Christer Björkman, Erika Iyengar, and Neal Williams for providing generous feedback on the first edition that greatly improved this second edition. We'd like to especially thank Truman Young for his extensive and detailed feedback on the first edition, which influenced our writing on practically every page of this edition. We are fairly certain that we have neglected to mention many others, and we apologize for these unintentional omissions. We are also grateful to Alison Kalett for really getting our vision for this book and supporting us in the process, and to Jodi Beder for unusually insightful editing—they both made writing the second edition of this book a pleasure.

References

Agrawal, A. A., and P. M. Kotanen. 2003. Herbivores and the success of exotic plants: A phylogenetically controlled experiment. *Ecology Letters* 6:712–715.

Baldwin, I. T. 1988. The alkaloidal responses of wild tobacco to real and simulated herbivory. *Oecologia* 77:378–381.

Beck, M. 2001. *Finding Your Own North Star.* Three Rivers Press, New York.

Bergerud, A. T., and W. E. Mercer. 1989. Caribou introductions in eastern North America. *Wildlife Society Bulletin* 17:111–120.

Blickley, J. L., K. Deiner, K. Garbach, I. Lacher, M. H. Meek, L. M. Porensky, M. L. Wilkerson, E. M. Winford, and M. W. Schwartz. 2013. Graduate student's guide to necessary skills for nonacademic conservation careers. *Conservation Biology* 27:24–34.

Bolles, R. N. 2013. *What Color Is Your Parachute?* Ten Speed Press, New York.

Bowman, J., J. C. Ray, A. J. Magoun, D. S. Johnson, and F. N. Dawson. 2010. Roads, logging, and the large-mammal community of an eastern Canadian boreal forest. *Canadian Journal of Zoology* 88:454–467.

Brown, J. H., and M. V. Lomolino. 1989. Independent discovery of the equilibrium theory of island biogeography. *Ecology* 70:1955–1957.

Burnham, K. P., and D. R. Anderson. 2002. *Model Selection and Multimodel Inference: A Practical Information-Theoretic Approach.* 2nd ed. Springer Verlag, New York.

Calcagno, V., E. Demoinet, K. Gollner, L. Guidi, D. Ruths, and C. de Mazancourt. 2012. Flows of research manuscripts among scientific journals reveal hidden submission patterns. *Science* 338: 1065–1069.

Cassey, P., and T. M. Blackburn. 2004. Publication and rejection among successful ecologists. *BioScience* 54:234–239.

Chandler, C. R., L. M. Wolfe, and D. E. L. Promislow. 2007. *The Chicago Guide to Landing a Job in Academic Biology.* University of Chicago Press, Chicago.

Cohen, J. 1988. *Statistical Power Analysis for the Behavioral Sciences.* 2nd ed. Lawrence Erlbaum, Hillsdale, NJ.

Colzato, L. S., A. Ozturk, and B. Hommel. 2012. Meditate to create: The impact of focused-attention and open-monitoring training on convergent and divergent thinking. *Frontiers in Psychology* 3:116.

Cottingham, K. L., J. T. Lennon, and B. L. Brown. 2005. Knowing when to draw the line: Designing more informative ecological experiments. *Frontiers in Ecology and the Environment* 3:145–152.

Crouse, D. T., L. B. Crowder, and H. Caswell. 1987. A stage-based population model for loggerhead sea turtles and implications for conservation. *Ecology* 68:1412–1423.

Damrosch, D. 1995. *We Scholars: Changing the Culture of the University.* Harvard University Press, Cambridge, MA.

Darwin, C. 1889. *The Origin of Species.* 6th ed. D. Appleton, New York.

Deegan, D. H. 1995. Exploring individual differences among novices reading in a specific domain: The case of law. *Reading Research Quarterly* 30:154–157.

Diamond, J. 1986. Overview: Laboratory experiments, field experiments, and natural experiments. Pages 3–22 in J. Diamond and T. J. Case (eds.), *Community Ecology.* Harper & Row, New York.

Dray, S., R. Pelissier, P. Couteron, M. J. Fortin, P. Legendre, P. R. Peres-Neto, E. Bellier, R. Bivand, F. G. Blanchet, M. De Caceres, A. G. Dufour, E. Heegaard, T. Jombart, F. Munoz, J. Oksanen, J. Thioulouse, and H. H. Wagner. 2012. Community ecology in the age of multivariate multiscale spatial analysis. *Ecological Monographs* 82:257–275.

Elzinga, C. L., D. W. Salzer, J. W. Willoughby, and J. P. Gibbs. 2001. *Monitoring Plant and Animal Populations.* Wiley, Hoboken, NJ.

Erren, T. C., and P. E. Bourne. 2007. Ten simple rules for a good poster presentation. *PLoS Computational Biology* 3(5):e102.

Feist, G. J. 1998. A meta-analysis of personality in scientific and artistic creativity. *Personality and Social Psychology Review* 2:290–309.

Felton, G. W., and H. Eichenseer. 1999. Herbivore saliva and its effects on plant defense against herbivores and pathogens. Pages 19–36 in A. A. Agrawal, S. Tuzin, and E. Bent (eds.), *Induced Plant Defenses against Pathogens and Herbivores: Biochemistry, Ecology, and Agriculture.* American Phytopathological Society Press, St. Paul, MN.

Finkbeiner, E. M., B. P. Wallace, J. E. Moore, R. L. Lewison, L. B. Crowder, and A. J. Read. 2011. Cumulative estimates of sea turtle bycatch and mortality in USA fisheries between 1990 and 2007. *Biological Conservation* 144:2719–2727.

Fleet, C. M., M.F.N. Rosser, R. A. Zufall, M. C. Pratt, T. S. Feldman and P. P. Lemons. 2006. Hiring criteria in biology departments of academic institutions. *BioScience* 56:430–436.

Foucault, M. 1977. *Discipline and Punish: The Birth of the Prison*. Pantheon Books, New York.

Friedland, A. J., and C. L. Folt. 2009. *Writing Successful Science Proposals*. 2nd ed. Yale University Press. New Haven, CT.

Futuyma, D. J. 1998. Wherefore and whither the naturalist? *American Naturalist* 151:1–6.

Garland, T., A. F. Bennett, and E. L. Rezende. 2005. Phylogenetic approaches in comparative physiology. *Journal of Experimental Biology* 208:3015–3035.

Gold, C. M., and T. M. Dore. 2001. At cross purposes: What the experiences of doctoral students reveal about doctoral education (www.phd-survey.org). A report prepared for the Pew Charitable Trusts, Philadelphia.

Gotelli, N. J., and A. M. Ellison. 2004. *A Primer of Ecological Statistics*. Sinauer, Sunderland, MA.

Grace, J. B. 2006. *Structural Equation Modeling and Natural Systems*. Cambridge University Press, Cambridge.

Gurevitch, J., and L. V. Hedges. 2001. Meta-analysis: Combining the results of independent experiments. Pages 347–369 in S. M. Scheiner and J. Gurevitch (eds.), *Design and Analysis of Ecological Experiments*, 2nd ed. Oxford University Press, Oxford, UK.

Hilborn, R., and M. Mangel. 1997. *The Ecological Detective: Confronting Models with Data*. Princeton University Press. Princeton, NJ.

Hofer, T., H. Przyrembel, and S. Verleger. 2004. New evidence for the Theory of the Stork. *Paediatric and Perinatal Epidemiology* 18: 88–92.

Holt, R. D. 1977. Predation, apparent competition and the structure of prey communities. *Theoretical Population Biology* 12:197–229.

Holt, R. D., and J. H. Lawton. 1994. The ecological consequences of shared natural enemies. *Annual Review of Ecology and Systematics* 25:495–520.

Huberty, A. F., and R. F. Denno. 2004. Plant water stress and its consequences for herbivorous insects: A new synthesis. *Ecology* 85: 1383–1398.

Huntzinger, M. 2003. Effects of fire management practices on but-
    terfly diversity in the forested western United States. *Biological
    Conservation* 113:1–12.

Huntzinger, M., R. Karban, T. P. Young, and T. M. Palmer. 2004. Re-
    laxation of induced indirect defenses of acacias following exclu-
    sion of mammalian herbivores. *Ecology* 85:609–614.

Hurlbert, S. H. 1984. Pseudoreplication and the design of ecological
    field experiments. *Ecological Monographs* 54:187–211.

Karban, R. 1983. Induced responses of cherry trees to periodical ci-
    cada oviposition. *Oecologia* 59:226–231.

Karban, R. 1987. Environmental conditions affecting the strength of
    induced resistance against mites in cotton. *Oecologia* 73:414–419.

Karban, R. 1989. Community organization of *Erigeron glaucus* foli-
    vores: Effects of competition, predation, and host plant. *Ecology*
    70:1028–1039.

Karban, R. 1993. Costs and benefits of induced resistance and plant
    density for a native shrub, *Gossypium thurberi. Ecology* 74:9–19.

Karban, R., and P. de Valpine. 2010. Population dynamics of an arc-
    tiid caterpillar-tachinid parasitoid system using state-space mod-
    els. *Journal of Animal Ecology* 79:650–661.

Karban, R., and J. Maron. 2001. The fitness consequences of inter-
    specific eavesdropping between plants. *Ecology* 83:1209–1213.

Karban, R., T. Mata, P. Grof-Tisza, G. Crutsinger, and M. Holyoak.
    2013. Non-trophic effects of litter reduce ant predation and deter-
    mine caterpillar survival and distribution. *Oikos* 122:1362–1370.

Kearns, C. A., and D. W. Inouye. 1993. *Techniques for Pollination Biolo-
    gists.* University Press of Colorado, Niwot, CO.

Koenig, W. D., and J.M.H. Knops. 2013. Large scale spatial syn-
    chrony and cross-synchrony in acorn production by two Califor-
    nia oaks. *Ecology* 94:83–93.

Koenig, W. D., J.M.H. Knops, and W. J. Carmen. 2010. Testing the
    environmental prediction hypothesis for mast-seeding in Califor-
    nia oaks. *Canadian Journal of Forest Research* 40:2115–2122.

Koenig, W. D., J.M.H. Knops, W. J. Carmen, M. T. Stanback, and
    R. L. Mumme. 1996. Acorn production by oaks in central coastal
    California: Influence of weather at three levels. *Canadian Journal
    of Forest Research/Revue Canadienne De Recherche Forestière* 26:1677–
    1683.

Legendre, P., M.R.T. Dale, M.-J. Fortin, F. Gurevitch, M. Hohn, and
    D. Myers. 2002. The consequences of spatial structure for the de-

sign and analysis of ecological field surveys. *Ecography* 25:601–615.

Lertzman, K. 1995. Notes on writing papers and theses. *Bulletin of the Ecological Society of America* June 1995:86–90.

MacArthur, R. H., and E. O. Wilson. 1963. An equilibrium theory of insular zoogeography. *Evolution* 17:373–387.

MacArthur, R. H., and E. O. Wilson. 1967. *The Theory of Island Biogeography*. Princeton University Press, Princeton, NJ.

Maron, J. L., and S. Harrison. 1997. Spatial pattern formation in an insect host-parasitoid system. *Science* 278:1619–1621.

Marquis, R. J., and C. J. Whelan. 1995. Insectivorous birds increase growth of white oak through consumption of leaf-chewing insects. *Ecology* 75:2007–2014.

Martinsen, O. L. 2011. The creative personality: A synthesis and development of the creative person profile. *Creativity Research Journal* 23:185–202.

Matthews, R. 2000. Storks deliver babies (p = 0.008). *Teaching Statistics* 22:36–38.

Mitchell, R. J. 2001. Path analysis: Pollination. Pages 217–234 in S. M. Scheiner and J. Gurevitch (eds.), *Design and Analysis of Ecological Experiments*, 2nd ed. Oxford University Press, Oxford, UK.

Monroe, E. G. 1948. The geographical distribution of butterflies in the West Indies. Ph.D. dissertation, Cornell University, Ithaca, NY.

Monroe, E. G. 1953. The size of island faunas. Pages 52–53 in *Proceedings of the Seventh Pacific Science Congress of the Pacific Science Association*. Volume 4: *Zoology*. Whitcome and Tombs, Auckland, New Zealand.

Moore, P. D., and S. B. Chapman. 1986. *Methods in Plant Ecology*. 2nd ed. Blackwell Scientific Publications, Oxford, UK.

Newmark, W. D. 1995. Extinction of mammal populations in western North American national parks. *Conservation Biology* 9:512–526.

Newmark, W. D. 1996. Insularization of Tanzanian parks and the local extinction of large mammals. *Conservation Biology* 10:1549–1556.

Oksanen, L. 2001. Logic of experiments in ecology: Is pseudoreplication a pseudoissue? *Oikos* 94:27–38.

Pearse, I. S. 2011. The role of leaf defensive traits in oaks on the preference and performance of a polyphagous herbivore, *Orgyia vetusta*. *Ecological Entomology* 36:635–642.

Pearse, I. S., and A. L. Hipp. 2009. Phylogenetic and trait similarity to a native species predict herbivory on non-native oaks. *Proceedings of the National Academy of Sciences* 106:18097–18102.

Pechmann, J.H.K., D. E. Scott, R. D. Semlitsch, J. P. Caldwell, L. J. Vitt, and J. W. Gibbons. 1991. Declining amphibian populations: The problem of separating human impacts from natural fluctuations. *Science* 253:892–895.

Platt, J. R. 1964. Strong inference. *Science* 146:347–353.

Popper, K. R. 1959. *The Logic of Scientific Discovery*. Basic Book, New York.

Potvin, C. 1993. ANOVA: Experiments in controlled environments. Pages 46–68 in S. M. Scheiner and J. Gurevitch (eds.), *Design and Analysis of Ecological Experiments*. Chapman and Hall, New York.

Quinn, J. F., and A. E. Dunham. 1983. On hypothesis testing in ecology and evolution. *American Naturalist* 122:602–617.

Reznick, D. N., and J. A. Endler. 1982. The impact of predation on life history evolution in Trinidadian guppies (*Poecilia reticulata*). *Evolution* 36:160–177.

Reznick, D. N., H. Bryga, and J. A. Endler. 1990. Experimentally induced life-history evolution in a natural population. *Nature* 346: 357–359.

Ricklefs, R. E. 2012. Naturalists, natural history, and the nature of biological diversity. *American Naturalist* 179:423–435.

Ricklefs, R. E., and D. Schluter. 1993. Species diversity: Regional and historical influences. Pages 350–363 in R. E. Ricklefs and D. Schluter (eds.), *Species Diversity in Ecological Communities*. University of Chicago Press, Chicago.

Schneider, D. C., R. Walters, S. Thrush, and P. Dayton. 1997. Scale-up of ecological experiments: Density variation in the mobile bivalve *Macomona liliana*. *Journal of Experimental Marine Biology and Ecology* 216:129–152.

Shipley, B. 2000. *Cause and Correlation in Biology: A User's Guide to Path Analysis, Structural Equations and Causal Inference*. Cambridge University Press, Cambridge, UK.

Shurin, J. B., E. T. Borer, E. W. Seabloom, K. Anderson, C. A. Blanchette, B. Broitman, S. D. Cooper, and B. S. Halpern. 2002. A cross-ecosystem comparison of the strength of trophic cascades. *Ecology Letters* 5:785–791.

Singer, M. S., T. E. Farkas, C. M. Skorik, and K. A. Mooney. 2012. Tritrophic interactions at a community level: Effects of host plant species quality on bird predation of caterpillars. *American Naturalist* 179:363–374.

Sokal, R. R., and F. J. Rohlf. 2012. *Biometry.* 4th ed. Freeman, New York.

Southwood, T.R.E., and P. A. Henderson. 2000. *Ecological Methods.* Blackwell Science, Oxford, UK.

Sutherland, W. J. (ed.). 2006. *Ecological Census Techniques.* 2nd ed. Cambridge University Press, Cambridge, UK.

Thomas, D. C., and D. R. Gray. 2002. Update COSEWIC status report on the woodland caribou *Rangifer tarandus caribou* in Canada. In *COSEWIC assessment and update status report on the woodland caribou* Rangifer tarandus caribou *in Canada.* Committee on the Status of Endandered Wildlife in Canada. Ottawa.

Thompson, J. N. 1999. Specific hypotheses on the geographic mosaic of coevolution. *American Naturalist* 153 Supplement:S1–S14.

Todd, B. T., D. E. Scott, J.H.K. Pechmann, and J. W. Gibbons. 2011. Climate change correlates with rapid delays and advancements in reproductive timing in an amphibian community. *Proceedings of the Royal Society B* 278:2191–2197.

Tyler, C. M., B. Kuhn, and F. W. Davis. 2006. Demography and recruitment limitation of three oak species in California. *Quarterly Review of Biology* 81:127–152.

Van Kammen, D. P. 1987. Columbus, grantsmanship, and clinical research. *Biological Psychology* 22:1301–1303.

Vaughn, K. J., and T. P. Young. 2010. Contingent conclusions: Year of initiation influences ecological field experiments, but temporal replication is rare. *Restoration Ecology* 18:59–64.

Vitt, L. J., and E. R. Pianka. 2005. Deep history impacts present-day ecology and biodiversity. *Proceedings of the National Academy of Sciences* 102:7877–7881.

Weber, M. G., and A. A. Agrawal. 2012. Phylogeny, ecology, and the coupling of comparative and experimental approaches. *Trends in Ecology and Evolution* 27:394–403.

White, T.C.R. 1969. An index to measure weather-induced stress of trees associated with outbreaks of psyllids in Australia. *Ecology* 50:905–909.

White, T.C.R. 1984. The abundance of invertebrate herbivores in relation to the availability of nitrogen in stressed food plants. *Oecologia* 63:90–105.

White, T.C.R. 2008. The role of food, weather and climate in limiting the abundance of animals. *Biological Reviews* 83:227–248.

Wilson, D. E., F. R. Cole, J. D. Nichols, R. Rudran, and M. S. Foster. 1996. *Measuring and Monitoring Biological Diversity: Standard Methods for Mammals.* Smithsonian Institution Press, Washington, DC.

Yang, L. H. 2004. Periodical cicadas as resource pulses in North American forests. *Science* 306:1565–1567.

Yoccuz, N. G. 1991. Use, overuse, and misuse of significance tests in evolutionary biology and ecology. *Bulletin of the Ecological Society of America* 72:106–111.

Young, T. P. 2000. Restoration ecology and conservation biology. *Biological Conservation* 92:73–83.

Young, T. P., B. Okello, D. Kinyua, and T. M. Palmer. 1998. KLEE: A long-term, large-scale herbivore exclusion experiment in Laikipia, Kenya. *African Journal of Range and Forage Science* 14:94–102.

Zschokke, S., and E. Ludin. 2001. Measurement accuracy: How much is necessary? *Bulletin of the Ecological Society of America* 82:237–243.

Index